# Lasers in Analytical
# Atomic Spectroscopy

# Lasers in Analytical Atomic Spectroscopy

*EDITED BY*

*Joseph Sneddon, Terry L. Thiem, and Yong-Ill Lee*

Joseph Sneddon
Department of Chemistry
McNeese State University
Lake Charles, LA 70609

Terry L. Thiem
Technical Assistance International
1320 Loma Verde
El Paso, TX 79936

Yong-Ill Lee
Department of Chemistry
Konyang University
Nonsan, Chungnam 320-800
The Republic of Korea

This book is printed on acid-free paper. ♾

**Library of Congress Cataloging-in-Publication Data**
Lasers in analytical atomic spectroscopy / edited by Joseph Sneddon,
    Terry L. Thiem, and Yong-Ill Lee.
        p.   cm.
    Includes bibliographical references and index.
    ISBN 1-56081-907-3 (acid-free paper)
    1. Atomic spectroscopy.   2. Lasers in chemistry.   I. Sneddon,
Joseph, 1951-    . II. Thiem, Terry L. III. Lee, Yong-Ill.
QD96.A8L38 1996
543′.0873—dc20                                                    96-14917
                                                                      CIP

© 1997 VCH Publishers, Inc.

ISBN 1-56081-907-3 VCH Publishers, Inc.

Printing History
10 9 8 7 6 5 4 3 2 1

Published jointly by

VCH Publishers, Inc.
333 7th Avenue
New York, New York 10001

VCH Verlagsgesellschaft mbH
P.O. Box 10 11 61
69451 Weinheim, Germany

VCH Publishers (UK) Ltd.
8 Wellington Court
Cambridge CB1 1HZ
United Kingdom

# Preface

The invention of the laser occurred around 1960, and in its early stages of development was often referred to as "an invention without an application." Later years have shown the error of this statement; the laser has become an integral and indispensable part of many analytical techniques and instrumentation. It has found widespread use in Raman, molecular, and fluorescence spectrometry, but analytical atomic spectroscopy was slow to capitalize on the unique properties of lasers. This was due, in part, to the fact that conventional atomic spectroscopy frequently fulfilled the needs of the user. As lower levels of metals in complex samples need to be determined, the laser has found an increasing use in analytical atomic spectroscopy. At the present time, however, it has not found widespread use. This book attempts to show the current status of the use of lasers in analytical atomic spectroscopy, and primarily, in analytical atomic spectroscopy with optical detection. This has been achieved throughout the book with the exception of the inductively coupled plasma mass spectrometry (ICP-MS) technique described in Chapter 4.

Chapter 1 is in the form of a brief introduction and is intended to provide the reader with some basic information and instrumentation on conventional atomic spectroscopy. Chapter 2 describes the properties, types, and principles of lasers. A nonmathematical approach to lasers is used in this chapter. The goal is to provide the reader with some background about lasers. Chapters 3 through 6 are the heart of the book, and describe the major application of the laser to analytical atomic spectroscopy in detail. This includes laser-excited atomic fluorescence spectrometry (LEAFS) in Chapter 3 from Bob Michel and his students at the University of Connecticut in Storrs, Connecticut; laser ablation for sample introduction in Chapter 4, particularly in inductively coupled plasma atomic emission spectrometry (ICP-AES) and ICP-MS by Lieselotte

Moenke-Blankenburg at Martin-Luther-University Halle-Wittenberg in Halle, Germany; laser-induced breakdown (emission) spectrometry (LIBS) in Chapter 5 from Yong-Ill Lee of Konyang University in Nonsan in the Republic of Korea, Kyuseok Song at the Korea Atomic Energy Institute on Taejon in the Republic of Korea, and Joseph Sneddon from McNeese State University in Lake Charles, Louisiana; and laser-enhanced ionization (LEI) spectrometry in Chapter 6 from Dave Butcher at Western Carolina University in Cullowhee, North Carolina. The editors consider these to be the four major areas where there is widespread use of lasers in analytical atomic spectroscopy. In Chapter 1 there is a brief description of the use of lasers in other areas of analytical atomic spectroscopy. While having some potential, the editors do not feel that they require a complete chapter at present.

This book is intended to introduce the reader to the use of lasers in analytical atomic spectroscopy. It can be used as a short course text or a stand-alone text for a graduate course.

The editors thank the contributors for their contributions and timely submissions. We also thank the reviewers for their timely reviews and helpful suggestions.

Yong-Ill Lee
Terry L. Thiem
Joseph Sneddon

# Contents

# Contributors

**David J. Butcher**
Department of Chemistry and
Physics
Western Carolina University
Cullowhee, NC 28723

**Xiandeng Hou**
Department of Chemistry
University of Connecticut
215 Glenbrook Road
Storrs, CT 06269-4060

**Yong-Ill Lee**
Department of Chemistry
Konyang University
Nonsan, Chungnam 320-800
The Republic of Korea

**Robert F. Lonardo**
Department of Chemistry
University of Connecticut
215 Glenbrook Road
Storrs, CT 06269-4060

**Robert G. Michel**
Department of Chemistry
University of Connecticut
215 Glennbrook Road
Storrs, CT 06269-4060

**Lieselotte Moenke-Blankenburg**
Martin-Luther-University Halle-
Wittenberg
Department of Chemistry
Institute of Analytical and
Environmental Chemistry
Weinbergweg 16
D-06120 Halle
Germany

**Joseph Sneddon**
Department of Chemistry
McNeese State University
Lake Charles, LA 70609

**Kyuseok Song**
Quantum Optics Laboratory
Korea Atomic Energy Research
Institute
Taejon, 305-600
The Republic of Korea

**Terry L. Thiem**
Technical Assistance International
El Paso, TX 79936

**Suh-Jen Jane Tsai**
Department of Applied Chemistry
Providence University
200 Chungchi Road
Shalu 43301, Taichung Hsien
Taiwan, R.O.C.

**Karl X. Yang**
Department of Chemistry
University of Connecticut
215 Glenbrook Road
Storrs, CT 06269-4060

**Jack X. Zhou**
Department of Chemistry
University of Connecticut
215 Glenbrook Road
Storrs, CT 06269-4060

# 1

# Analytical Atomic Spectroscopy

## *Joseph Sneddon, Terry L. Thiem, and Yong-Ill Lee*

## 1.1 Introduction

The interaction of energy with matter produces three closely related, yet separate, phenomena: atomic absorption spectrometry (AAS), atomic emission spectrometry (AES), and atomic fluorescence spectrometry (AFS). These techniques
are collectively known as atomic spectroscopy. A closely related technique is
that of plasma source mass spectroscopy, in particular inductively coupled
plasma mass spectrometry (ICP-MS). Over the last several decades, these techniques have been applied to the quantitative, semiquantitative, and qualitative
determination of metals or elements in just about every conceivable type of
sample including agricultural, biochemical, clinical, chemical, environmental,
food, geology, industrial, and pharmaceutical. This includes solids, solutions,
and to a lesser extent, gaseous samples. The techniques are used primarily in
the quantitative mode for trace (low microgram per milliliter) and ultratrace
(microgram per liter and sub–microgram per liter) determinations.

This chapter presents a very brief and general overview of atomic spectroscopy as an analytical technique. The objective is to provide the reader with the
basic principles, instrumentation, and performance of conventional analytical
atomic spectroscopy. The remainder of the book (except Chapter 2 which gives
a description of the laser) will describe the application of the laser in analytical
atomic spectroscopy. More comprehensive and recent books devoted to the
general area of conventional atomic spectroscopy are available.[1–3] Individual
chapters in most instrumental, quantitative, or analytical chemistry textbooks

are devoted to these techniques.[4-6] The area of atomic spectroscopy is large, and more specific areas within atomic spectroscopy are covered in specialized books.[7-10] Several journals and book series present reviews or are primarily devoted to atomic spectroscopy.[11-15]

## 1.2  Historical

Atomic spectroscopy is considered to have started with a description of the visible spectrum in 1666 by Isaac Newton. In 1802, Wollaston[16] reported that our sun's spectrum contained a number of dark lines or bands. In 1823, Fraunhofer[17] measured the wavelengths and determined the origin of these lines. In 1859, Kirchoff[18] determined that these lines were caused by the absorption of atomic vapors. During this same period, Bunsen and Kirchoff[19] noted that the colors imparted to a flame by metallic salts were quite specific, and they thereby discovered atomic emission. Atomic fluorescence was not discovered until the beginning of the twentieth century by Wood.[20] From the mid–nineteenth century until the late 1920s, atomic spectroscopy was more closely associated with atomic physics. In 1928, Lundegrah demonstrated atomic emission in an air-acetylene flame and applied his system to agricultural analysis. Atomic emission spectrometry enjoyed considerable development and popularity in the analysis area at this time until the development of AAS independently in the mid-1950s by Walsh[21] and Alkemade and Melatz.[22] From that time to the present, AAS has rapidly grown to be widely accepted and become an almost indispensable technique for elemental determination in the laboratory. After the discovery of AAS, the use and development of AES decreased until the mid-1970s with the development of plasma, in particular inductively coupled plasma (ICP), as an excitation source. AES has become increasingly competitive with AAS primarily due to the multi-element analysis capability associated with the higher source temperature. More recently, beginning in the mid-1980s, ICP has been used as an ion source for mass spectroscopy (MS). The use of ICP-MS for trace and ultratrace metal determination is considered by many to be the future of atomic spectroscopy. The use of AFS as an analytical technique was developed independently in the early 1960s by Alkemade,[23] Winefordner,[24,25] and West.[26]

## 1.3  General Characteristics of Atomic Spectra

Spectroscopy is a science of the interaction of electromagnetic radiation with matter. The atomic absorption spectra for most metals originate from the transition of atoms from the ground state to one or more excited states. Atomic emissions originate when atoms are excited to a higher energy state or states and spontaneously emit their energy to arrive at lower or ground-state levels. Atomic fluorescence spectra originate when atoms at ground state are excited to a higher energy level and then spontaneously emit or fluoresce back to a lower ground state. A detailed discussion of atomic spectra is beyond the scope of this chapter and the reader is referred to other texts.[27,28]

## 1.4 Atomic Absorption Spectrometry

Atomic absorption involves the impingement of light of a specific wavelength, usually from a line source, on ground-state atoms. The atoms absorb the light and are excited to a higher energy level. The intensity of this transition is related to the concentration of the atoms in the ground state as follows:

$$T = I/I_0 \qquad (1.1)$$

where $T$ is the transmittance, $I$ is the intensity of the light source passing through the sample zone, and $I_0$ is the intensity of the light source before it passes through the sample zone. In AAS, the sample zone is relatively long to maximize the amount of light absorbed by the atoms. The presence of atoms of interest (selected using a monochromator) will attenuate $I$ and result in a positive absorbance. The presence of no atoms of interest should result in no absorbance signal. The amount of light absorbed will depend on the atomic absorption coefficient, $k$. This value is related to (1) the number of atoms per cubic centimeter in the atom cell, $n$; (2) the Einstein probability for the absorption process; and (3) the energy difference between the two levels of the transition. In practice, these are all constants that are combined to give one constant called the absorptivity, $a$. The coefficient $k$ is related exponentially to the transmittance as follows:

$$T = I/I_0 = e^{-kb} \qquad (1.2)$$

where $b$ is the path length or sample zone. In practice, the absorbance, $A$, is used in AAS and is related logarithmetically to the transmittance as follows:

$$A = -\log T = -\log I/I_0 = \log I_0/I$$
$$= \log 1/T = kb \log e = 0.43 \, kb \qquad (1.3)$$

The Beer-Lambert Law relates $A$ to the concentration of the element in the atom cell, $c$, as follows:

$$A = abc \text{ or } A = \varepsilon_0 bc \qquad (1.4)$$

where $a$ is the absorptivity in grams per liter centimeter, $\varepsilon_0$ is the molar absorptivity in grams per mole centimeter, and $b$ is the sample zone or cell width in centimeters. AAS involves the measurement of the drop in light intensity of $I_0$ to $I$ (depending on the concentration of the element). Current and modern instrumentation automatically converts the logarithmic value into $A$. Absorbance is a unitless number, typically 0.01 to 2.0. In practice, it is better to work in the middle of this range (0.1–0.3 is recommended) as the precision is reduced at the extremes due to instrumental noise. Also, the linearity AAS is reduced at around 0.3 and above. The most intense transition, from the ground state to the first excited state (resonance transition), is the most widely used transition because it is the most sensitive.

### 1.4.1 Instrumentation

The primary components of an AAS instrument are shown schematically in Figure 1.1 and consist of a radiation source, atomizer, sample introduction system, wavelength selection device, detector, amplifier, and readout system.

### 1.4.1.1 *Radiation Source*

The most widely used and accepted radiation source for AAS is the line source, in particular the hollow cathode lamp (HCL) and to a lesser extent the radio-frequency electrodeless discharge lamp (EDL). For the most part, the HCL is a line source, and each new metal to be tested for requires a separate lamp. Some multi-element lamps are available, for example, Ca-Mg and Cr-Fe-Ni, in which the cathode is made of two or three metals with similar properties. In general, these lamps produce a poorer performance when compared to the single-element lamp because of differences in volatilities that necessitate a compromise in operating conditions, but they do offer the possibility of limited multi-element determination. The cost of multi-element lamps is lower compared to the cost of combining individual lamps.

#### 1.4.1.1.a Hollow Cathode Lamp

A schematic diagram of a typical single-element HCL is shown in Figure 1.2. An HCL is a glass envelope filled with inert gas (usually neon, argon, or occasionally helium) at a pressure of 1–5 torr, and a hollow cathode made of a pure metal. About 500 V and a current of 2–30 mA is applied between the anode (positive) and the cathode (negative), of and the discharge concentrates in the hollow cathode. The filler gas is ionized (becomes charged) at the anode, and the positive ions produced (e.g., $Ar + e^- = Ar^+ + e^- + e^-$) are accelerated by the charge toward the negative cathode. The ions will impinge on or strike

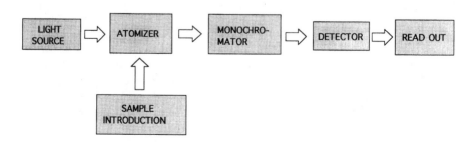

**Figure 1.1** Basic components of an atomic absorption spectrometer.

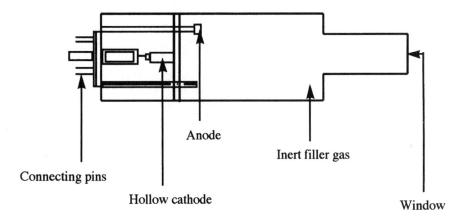

**Figure 1.2**   Schematic diagram of a typical hollow cathode lamp.

the cathode, causing the metal ions to be removed or "sputtered" out of the cathode. Further collisions excite the metal atoms, and the excited metal ions produce an intense, characteristic spectrum of the metal of interest when they return to the ground state. These atoms diffuse back to the cathode or onto the glass walls of the lamp. Modern lamps are constructed of glass, but require a silica window for use in the ultraviolet region, and have a molded plastic base. The seal between the window and the envelope must be gas tight to prevent loss of pressure and fill gas. HCLs do not have an infinite lifetime; typically, they last 2–5 A h, and can become redundant if not used. This is particularly true for elements with high volatility, such as arsenic and selenium. A short warm-up period of a few minutes is usually adequate prior to use. Current HCLs are very stable over several hours of continuous use.

### 1.4.1.1.b  Electrodeless Discharge Lamp

The EDL was originally developed for AFS and was excited primarily by microwaves. It is much more intense than the HCL but somewhat less stable. It is described briefly in the section on AFS instrumentation. Radiofrequency (RF) excited EDLs can be used in AAS. They are typically 5–100 times more intense than HCLs and operate in the 100 KHz to 100 Mz region. A commercially available RF-EDL is shown in Figure 1.3. Commercial RF-EDLs have a built-in starter, are pretuned, and are enclosed to stabilize the temperature. They run at 27.12 MHz at room temperature with a simple power supply. In AAS, the increased intensity does not lead to a significant improvement in sensitivity, and RF-EDLs are recommended for metals that do not make good HCLs. Such metals have low volatility and resonance lines in the low ultraviolet region of the electromagnetic spectrum. Examples include arsenic (193.7 nm), selenium (196.0 nm), zinc (213.9 nm), and cadmium (228.8 nm). RF-EDLs provide some

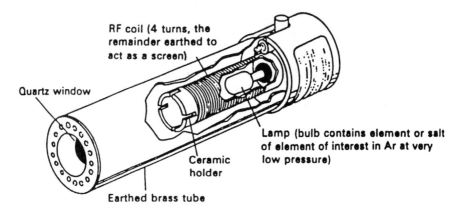

RF coil (4 turns, the
remainder earthed to
act as a screen)

Quartz window

Lamp (bulb contains element or salt
of element of interest in Ar at very
low pressure)

Ceramic
holder

Earthed brass tube

**Figure 1.3**   Schematic diagram of a radiofrequency electrodeless discharge lamp.

improvement in sensitivity in AAS (usually 1–3 times compared to the HCL), but are more expensive and do require the increased cost of the power supply.

### 1.4.1.1.c  Continuum Source

The use of a continuum source has been proposed for simultaneous multi-element AAS; one source could potentially replace up to 60 line sources. The most widely used continuum source is the high-intensity xenon arc lamp. It is most frequently coupled with an echelle monochromator or polychromator to obtain a spectral bandpass on the order of the atomic absorption linewidth. A bank of photo-multiplier tubes is placed behind appropriate exit slits for simultaneous analysis, or a scanning system may be used for sequential analysis.

While potentially very attractive, continuum source AAS does suffer several disadvantages when compared to line source AAS, such as poorer detection limits (a factor of between 2 and 4, in particular at resonance wavelengths below 280 nm) and nonlinearity or reduced linearity at high absorbances. This has been attributed to the fact that the spectral bandpass of the continuum source is still greater than the HCL line profile widths. Continuum source AAS is not available commercially. Much of the development, characterization, and applications have been achieved by O'Haver, Harnley, and coworkers.[29,30]

### 1.4.1.1.d  Other Sources

Various other sources have been proposed for AAS, including low-pressure discharge lamps, thermal gradient lamps, and plasmas. These sources have been proposed and evaluated for specific applications and have not been widely adopted or used. A new and potentially very exciting source is the tunable semiconductor diode laser, which is described in chapter 2. Its application in AAS is briefly discussed later in this chapter.

### 1.4.1.2 *Atomizers*

The object of the atomizer is to reduce the sample to an atomic vapor in a reproducible manner. This can be difficult because there are a number of dynamic processes that can occur in this step. In general, atomizers are a very inefficient means of producing atoms, but high sensitivities for quantitative detection can be obtained. The most widely used atomizer in AAS is the flame. The electrothermal atomizer or graphite furnace can be used for increased sensitivity or if there is a limited volume or mass of sample. These are the only two methods that have been widely adopted for producing atoms for AAS.

#### 1.4.1.2.a Flame

The structure of a premixed flame is shown in Figure 1.4. The premixed gases are heated in a preheating zone in which the temperature is increased exponentially until it reaches ignition temperature. Surrounding the preheating zone is

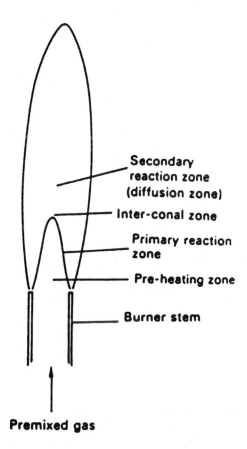

**Figure 1.4** The structure of a premixed flame.

the primary reaction zone where the most energetic reactions take place. The primary reaction zone is thin, typically around 1 mm, and thermodynamic equilibrium is not established. The partially combusted gases and flame radicals, such as OH·, CH·, and so forth, move through the flame and into the interconal zone. This is the hottest and most analytically useful part of the flame. At this stage, the hot gases come into contact with oxygen from the air, and products are formed in the secondary reaction zone. The premixed flame is the only system widely used in flame AAS, although some earlier work suggested that the laminar flame could be used in certain instances and applications.

The temperature will vary from one part of the flame to another. The air-acetylene flame, in which air is the oxidant and acetylene is the fuel, is the most widely used. It has an approximate temperature of 2400 K. To increase atomization efficiency, the nitrous oxide–acetylene flame, in which nitrous oxide is the oxidant and acetylene is the fuel, is most frequently used. It has a temperature of 3100 K. The flame fuel can be adjusted to produce a lean reducing flame through stoichiometric mixture to fuel rich conditions to optimize the absorbance signal from a particular element.

### 1.4.1.2.b  Electrothermal Atomizer

Electrothermal atomizers (ETAs) and the graphite furnace (GF) for AAS were developed in the late 1960s. Their principles advantage over flame atomizers are the improvement in sensitivity (typically 10–100 times), the ability to use microvolume (2–200 µL) and micromass (a few milligrams) sampling, and in situ pretreatment of the sample. However, ETAs are prone to interferences, particularly from alkali and alkaline earth halides, and require a more complex (and subsequently more expensive) system.

The use of an electrically heated tubular furnace was first reported by King[31] in 1905. However, for analytical chemistry, the work and system developed by L'vov[32,33] around 1960 is regarded as the forerunner of present-day ETA. This system is shown in Figure 1.5. It consisted of a carbon electrode in which the sample was placed, and a carbon tube that could be heated by electrical resistance. In the initial design, a supplementary electrode was used for preheating the furnace, the tube was lined with tungsten or tantalum foil to minimize vapor diffusion, and the system was purged with argon to prevent oxidation of the carbon. Later work involved direct heating of the sampling electrode by resistance heating, and the tube was made of pyrolitic carbon. After heating the tube, the sample electrode was inserted into the underside of the tube so that vaporization of the sample was confined to the tube, which is where AAS measurements were made. The system was difficult to operate and reproducibility was variable.

In 1967, Massmann[34] described a heated graphite atomizer (HGA) which was commercially developed by the Perkin-Elmer Corporation (Norwalk, Connecticut) and proved to be the forerunner for all current commercial ETAs. An isothermal-type furnace system proposed by Woodriff[35] at around the same time

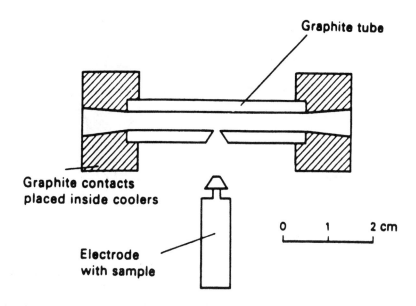

**Figure 1.5** The L'vov electrothermal atomizer.

was considered more difficult to commercialize, although recent work has shown the advantage of atomization under isothermal conditions. The Massmann system was a graphite tube, typically 50 mm long and 10 mm in diameter, which was heated by electrical resistance, typically 7–10 V at 400 A. An inert gas, usually argon or nitrogen, flowed through the atomizer at a constant rate (approximately 1.5 L/min), and the entire system was enclosed in a water jacket. The microliter sample was deposited through an entry or injection port in the center of the tube and, by applying variable current to the system, could be heated in three stages: drying to remove the solvent, ashing or pyrolysis to remove the matrix, and finally, atomization of the element. Careful control of the temperature was required in order to obtain good reproducibility.

In 1969, West and coworkers[36,37] developed a rod or filament atomizer. It consisted of a graphite filament 40 mm in length and 2 mm in diameter, supported by water-cooled electrodes, and heated very quickly with a current of 70 A at 10–12 V. Shielding from the air was achieved by a flow of inert gas around the filament. While primarily developed for AAS, West and coworkers showed the potential of the system for AFS.

The West filament was the forerunner of the mini-Massmann atomizer developed commercially by Varian Associates (Palo Alto, California). A commercial system called the carbon rod atomizer (CRA 63) is shown in Figure 1.6. Its main advantages were its somewhat simpler design compared to the HGA, low power requirements (2–3 kW), and fast (approximately 2 s) heating rate. There were differences between this system and the West filament, principally by drilling a hole in a solid cylindrical graphite tube and later using a small cup

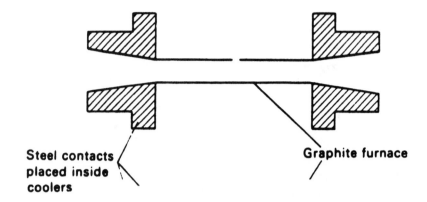

**Figure 1.6**   A carbon rod atomizer.

or crucible between two spring loaded graphite rods. The system was proposed for low-microliter volumes, typically 1–20 μL. Compared to the HGA, the CRA system had problems with detection limits and increased interference. This type of system was discontinued in the mid-1980s and is not currently commercially available.

### 1.4.1.2.c  Other Atomizers

Plasma has been evaluated as an atomization source for AAS. The direct current plasma (DCP) has been investigated by Wirz and coworkers.[38] Ng et al.[39] and Lu et al.[40] evaluated the microwave-induced plasma (MIP) as a source for AAS. Several authors have investigated the inductively coupled plasma (ICP) as an atomization source. Liang and Blades[41] developed and evaluated a capacitively coupled plasma (CCP).

While these plasmas have been shown to be useful for selected and specific applications, they have not and most probably will not be widely adopted as atomization sources in AAS. This is due to several factors including cost, complexity, and reduced analytical performance compared to conventional sources. The reduced performance can be attributed to smaller path length and the increased temperatures ($> 5000$ K) of the plasmas, which will produce more ions and fewer atoms.

### 1.4.1.3  Sample Introduction

The sample introduction system should reproducibly and efficiently transfer a sample to the atomizer. It should produce no interferences, be reproducible over extended times, have no memory or carryover effects, be independent of the sample type, and be universal for all types of atomizers or atomic spectroscopic techniques. Unfortunately all these desirable properties cannot be simultaneously obtained, so a compromise is necessary. Other factors to be considered when choosing a system include the amount of sample available (which may dictate a discrete or continuous delivery mode), the physical form of the sample (solid, liquid, slurry, or gas), the analytical performance characteristics desirable (precision, accuracy, detection limit, etc.), throughput of samples, and the type of atomizer (flame, furnace, or plasma). The sample introduction system should be viewed in conjunction with the atomization process. Sample introduction in atomic spectroscopy has been discussed effectively in a book edited by Sneddon.[8] Also, the work of Browner and coworkers[42,43] at the Georgia Institute of Technology in Atlanta, Georgia has contributed to our understanding of the sample introduction process in analytical atomic spectroscopy.

### 1.4.1.3.a Pneumatic Nebulizers

Pneumatic nebulizers (PN) are the most widely and commonly used method of introducing a solution. A jet of compressed air, the nebulization gas, aspirates and nebulizes the solution as the sample is sucked through a capillary tube. In the concentric nebulizer, the sample is surrounded by the oxidant gas as it emerges from the capillary tube, causing a reduced pressure at the tip and, thus, suction of the sample solution from the container (Bernoulli effect). In the angular or cross flow, a flow of compressed gas over the same capillary at right angles produces the same Bernoulli effect and aspirates the sample. The high velocity of this sample as it exits from the capillary tube creates a pressure drop and shatters the solution into tiny droplets. This aerosol then passes through a plastic expansion chamber where large droplets may be removed by impactors, spoilers, or baffles. This chamber also allows for the mixing of gases and has a drain tube to remove excess aerosol. A typical example of the chamber in combination with the pneumatic nebulizer and burner is shown in Figure 1.7. The object of this combination is to obtain a reproducible small aerosol diameter (typically 2–6 μm in size) prior to transport to the burner. In general, the pneumatic nebulizer does not have a very high nebulization efficiency (typically 5–10%) and is not recommended for high solid or salt solutions because of the potential for clogging and blockage of the capillary tubes.

### 1.4.1.3.b Other Nebulizers

Several other nebulization systems for the introduction of solutions to atomizers have been developed and characterized. They have been proposed to improve aerosol production (nebulization efficiency) and handle solutions with a high content of dissolved solids, more efficiently.

The ultrasonic nebulizer generates a homogeneous aerosol of fine (< 5 μm) droplets, has a high efficiency (30% or more), and can allow the carrier gas flow rate and aerosol generation rate to be independently variable. The basic principle involves the use of ultrasound waves (typically 0.1–10 MHz) to create discrete droplets from a solution, which are then swept to the atomizer through a chamber. Numerous workers[44,45] have investigated the ultrasonic nebulizer's

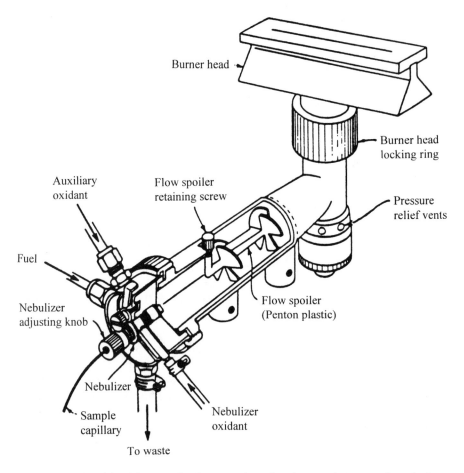

**Figure 1.7** Combined burner head, expansion chamber, and pneumatic nebulizer. (Courtesy of Perkin-Elmer Corp.)

principles, piezoelectric transducer, and design, as well as its application to difficult samples such as seawater and other high-salt solutions.

The Babington nebulizer has been used for solutions with a wide viscosity range, including high salt solutions and those with suspended particles. The basic principle involves allowing the solution to be nebulized to flow over a glass sphere that contains a small hole, slot, or series of very fine holes. Gas forced through these holes disrupts the the solution flowing over the sphere and produces an aerosol. Denton and coworkers[46,47] have developed this system and applied it to tomato sauce, evaporated milk, and untreated whole blood.

Various other proposed nebulizers include the thermospray, frit, and high solids nebulizers. In general, they are recommended for specific samples and are not universally used or widely accepted.

### 1.4.1.3.c  Other Sample Introduction Systems

Various sample introduction systems have been proposed including electro-thermal vaporization (ETV), laser ablation, direct insertion of solids, spark and arc, flow injection analysis, hydride generation, impaction, cold-vapor mercury, and low-pressure discharges.

## 1.4.1.4  Wavelength Selection Devices

About the only wavelength selection device used in AAS is the monochromator whose primary function is to isolate the wavelength of interest from other wave-lengths. A monochromator consists of two slits that provide an entrance and an exit, and a dispersing unit that consists of a prism or grating, plus some lenses or mirrors. The image of the entrance slit is transferred to the exit slit after dispersion of the wavelength components. An example of a commonly used monochromator, the Czerney-Turner grating monochromator, is shown in Figure 1.8. AAS has few spectral interferences due to the element-specific line source used, and the resolving power of the monochromator is not as critical as it is for AES. The monochromator should be capable of separating two wavelengths 0.2 nm apart when operating at a minimum effective slit width. The resolution, R, is dependent on the size and dispersing characteristics of the grating and the slit width of the monochromator, and is obtained by

$$R = \lambda/d\lambda = \omega \cdot d\theta/d\lambda \qquad (1.5)$$

where $\omega$ is the effective aperture width, $\lambda$ is the wavelength of the line to be measured, and $d\theta/d\lambda$ is the angular dispersion. The spectral bandpass is the range of radiation that passes through the exit slit. An optimum slit width is the spectral bandpass at which the wavelength of interest reaches the detector.

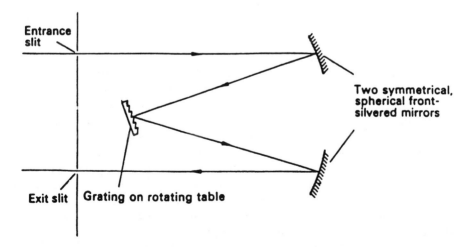

**Figure 1.8**  Schematic diagram of Czerney-Turner grating monochromater.

The diffraction grating is used to disperse the radiation into individual wavelengths. The dispersion, $dn/d\lambda$, depends on the refractive index of the prism material, n, and the wavelength, $\lambda$, of the line to be measured. The dispersion is high in the ultraviolet region but decreases sharply with increasing wavelength. The reciprocal linear dispersion, $d\lambda/dx$, is a function of the geometric slit width, S, and spectral bandpass, $\Delta\lambda_m$, of the monochromator such that

$$d\lambda/dx = \Delta\lambda_m/S \qquad (1.6)$$

and is usually expressed in nanometers per millimeter. If the desired bandpass is 0.2 nm and the reciprocal linear dispersion is 2 nm/mm, a geometric slit width of 0.1 mm must be obtained.

The reciprocal linear dispersion for a grating is nearly constant over the entire wavelength region and is dependent on the number of grooves per unit width, spectral order, and the focal length of the collimator. The resolution of a grating is a function of the spectral order, $m$, and the total number of grooves, $N$, such that

$$R = mN \qquad (1.7)$$

The effective aperture width of a grating is the width of an individual groove, d, multiplied by the total number of grooves, N, and by cos r (r is the angle of reflection) such that

$$\omega = dN \cos r \qquad (1.8)$$

If a grating ruled with 2000 grooves/nm and 50 mm in width has a resolution in the first order (AAS measurements are almost always done in the first order) of 100,000, at the sodium wavelength of 589 nm, the smallest wavelength interval resolved will be $\Delta\lambda = 589/100,000 = 0.006$ nm. Ratings of 2000–3000 grooves/nm and reciprocal dispersion of 1 nm/mm are usually considered good. Ruled gratings are used mainly in AAS.

### 1.4.1.5 Detectors

At present, the only detector widely used in AAS is the photomultiplier tube (PMT). Its purpose is to convert the radiation or light from the atomizer into an electrical signal. A typical PMT consists of a photosensitive cathode and a collection anode. These two electrodes are separated by several dynodes. The purpose of including one or more dynodes is to provide electron multiplication, and these are maintained at a potential between the two electrodes. The cathode is maintained at a negative 400–2500 V compared to the anode. The photoelectron ejected by the cathode will be attracted to the first dynode and eject several other electrons, which will then proceed to the next dynode. A cascade effect will result and a final gain, after moving through all the dynodes to the anode, of around $1 \times 10^6$ can be routinely achieved.

The popularity of PMTs can be attributed to their ease of operation, high sensitivity, reliability, and modest cost. An ultraviolet-sensitive PMT can effectively cover the wavelength range from just below 200 to nearly 600 nm with high sensitivity. This includes most of the resonance wavelengths used in AAS. However, a few of the alkali elements, such as lithium at 670 nm, and alkaline earth elements, such as potassium at 776 nm, require a red-sensitive PMT for high sensitivity. A drawback of PMTs is that they do not easily lend themselves to multichannel detection because of their size. If multichannel detection is desired, then a separate PMT is usually required for each element to be detected.

### *1.4.1.6 Readout*

The electrical signal from the detector is converted into absorbance units by a logarithmic amplifier. GF-AAS requires fast electronics to detect and measure the transient signal. The readout systems in AAS are closely linked with signal processing. This can include various operations such as converting current to voltage, analog data to digitaleem, and so forth. Readout systems can vary from system to system but some form of digital or graphic display is included. Current and cutting edge instrumentation in AAS usually involves computer hardware. This facilitates readout and instrument control of factors such as the experimental program in GF-AAS.

## 1.5  Recent Advances in AAS

### 1.5.1  Recent Developments in ET-AAS

The flame used in AAS has, for the most part, been unchanged over the last 20 years with improvements confined to safety and software. The last 20 years has seen tremendous improvements and understanding in electrothermal (ET) atomization for AAS. The subject and advances are worthy of a book devoted to the topic, and major advances are only briefly discussed here.

The most widely used material is graphite or carbon. A recent article[48] discusses the role of the atomization surface in ET-AAS. Graphite has many desirable properties as an electrothermal atomizer including very good isotropic and thermal conductivity, high stability up to 3300 K, low cost, easy machinability, high purity, and relatively good resistance to oxidation. However, it does suffer from some disadvantages including its porous nature which results in permeability to hot vapor; diffusional losses through a tube wall can be comparable to losses through the injection port. In addition, the graphite surface can be degraded by corrosive acids, and sample components, such as vanadium and molybdenum, may react to form carbides which are difficult to remove or atomize at the temperature (around 3300 K) attainable by a graphite furnace and can result in a type of interference called a memory effect. For these reasons, atomizers or atomization surfaces such as tungsten, platinum, tantalum, and rhenium inserted into graphite tubes have been proposed and evaluated. Several grades of graphite, such as pyrolytic coating carbon, glassy carbon, and totally pyrolytic carbon, have been developed to overcome some of the problems.

As stated above, the major drawback to ET-AAS (except time of analysis) is the increased interferences which can significantly reduce accuracy. Several methods have been proposed to minimize these interferences, including the use of matrix or chemical modifiers, platform atomization, background correction devices, and improved tube and system design. It is generally accepted that a combination of these improvements will lead to reduction or elimination of interferences and improved accuracy.

Background absorption due primarily to molecular absorption and light scattering effects can be a major interference in ET-AAS, particularly in the low ultraviolet region of the electromagnetic spectrum. A detailed study of the origin of interfering background absorption is available.[49] Several methods have been proposed to reduce this interference, and a review article is available on background correction in atomic spectroscopy.[50] These include traditional methods such as the use of a continuum source, and newer methods such as the Zeeman method and the Smith-Hieftje proposal to use pulsed hollow cathode lamps.

In recent years, the need for multi-element determinations has grown and AAS, which has mostly been regarded and used as a single element technique, has responded to this challenge. Multi-element AAS, for the most part, is limited to two to six elements and involves the used of more than one PMT. Thermo Jarrell Ash-Baird (Franklin, Massachusetts) have a unique system which uses one PMT but allows up to eight elements (in two batches of four elements) to be measured in a near simultaneous manner. This is accomplished through the use of a synchronized, rapidly moving, galvanometer-driven grating and lamp selection mirror, which allows for both multi-element and multiwavelength analysis. The spectrum from 190 to 900 nm is scanned in 20 ms. Multi-element AAS has been reviewed by Farah and Sneddon.[51]

### 1.5.2 Lasers Used as Light Sources in AAS

#### 1.5.2.1 Dye Lasers for AAS

New applications for lasers as light sources in atomic spectroscopy are continually being found and refined, resulting in improved accuracy and lower detection limits for many types of analysis. Dye lasers have been the tunable workhorse for both atomic and molecular spectroscopy. Although a variety of lasers have been used as pump sources (Nd:YAG, ruby, $N_2$, excimer, etc.), the range of wavelength tunability of the laser light is the key feature. Tunability of the laser light ranges from 220 to 800 nm. The most popular dye is the rhodamine 6G dye which allows tuning from 570 to 600 nm or, frequency doubled by second harmonic generation, from 285 to 300 nm.

Spectral interferences from background absorption and scattering are well known in flame atomic absorption spectroscopy. Background absorption may be caused by molecular absorption from CH and $C_2$ molecules present in the flame. To correct for these interferences, several different methods have been applied, including Zeeman background correction which uses magnetic fields to split electronic levels, Smith-Heifje background correction which uses pulsed hollow cathode lamps, continuous source compensation using a $D_2$ arc, two light sources (one of the element of interest and the other producing a line very close to that of the analyte of interest), and a tunable laser source such as a dye laser.

Kuhl et al.[52] studied the absorption of sodium in an analytical flame. The tunability of the dye laser permitted the scanning of wavelengths in the vicinity of the absorption line within a spectral range of 3 Å (0.3 nm) by tilting a Farby-Perot interferometer. The scanning of wavelength allowed background correction to be done almost simultaneously during the measurement. Using this method they were able to achieve a detection limit of 2 ng/cm$^{-3}$.

### 1.5.2.2 Semiconductor Diode Lasers for AAS

One of the hottest areas in laser technology research is the semiconductor diode laser. Although diode lasers were first used in atomic spectroscopy in the 1960s, they did not see widespread use until the 1980s, when room temperature diode lasers became commercially available. Worldwide, more than one million laser diodes are produced every month for use in compact disk players, bar code readers, and laser printers. Diode lasers help carry communications data through fiber optic cables. Diode lasers are efficient, inexpensive, compact, and tunable over a limited wavelength range, usually 20 nm. In addition, they have a long life ($10^5$ h), a high conversion efficiency from electricity to light, and can be directly driven by simple microelectronics. Unfortunately, diode lasers are not ideal spectroscopic sources because they have poor beam profiles caused by diffraction at the emitting aperture and are not continuously tunable due to mode hopping. Furthermore, the wavelength is restricted to the near infrared and red regions of the electromagnetic spectrum. This is a result of an energy gap of the semiconductor used.

Traditionally, lasers used for atomic spectroscopy have been tunable dye lasers. The problem with these systems is that they are very expensive, some costing $50,000 or more. It is now possible to purchase a diode laser system that will produce more than 10 mW of tunable light with a bandwidth of 100 kHz for less than $1000.[53] A virtue of the diode laser is that its amplitude is very stable compared to most other laser sources, so it is relatively simple to make sensitive absorption or fluorescence measurements. Diode lasers have already been used for many different applications in atomic spectroscopy. Replacing the traditional hollow cathode lamps (HCLs) used in most atomic absorption instruments with diode lasers would allow the user to perform background-corrected multi-element AAS with high selectivity, a large dynamic range, and internal standardization. If we look at a typical atomic absorption instrument (see figure 1.1), we see that the HCL is essential for a single element analysis.

To date, diode lasers have been used primarily for measuring the absorption of alkali metals such as Rb,[54] K, Sr, Ba, Cs, and Li. Rubidium was the first element studied at 780.02 nm in an air-hydrogen flame because this wavelength was achievable using the primary wavelengths of the first room temperature diode lasers. As the range of available wavelengths increased, so did the number of elements for which resonant lines could be probed using the laser. Simultaneous elemental analysis of Rb and Ba was demonstrated,[55] followed closely by the analysis of up to six elements with a grid of diode lasers.[56] These researchers[56] also found that using the wings of the absorption line extended the linear range of the calibration curve for Li. The detection limits realized using this method are found in Table 1.1.

## 1.6 Atomic Emission Spectrometry

Atomic emission spectrometry (AES) involves the impingement of an external source of energy on ground-state atoms. The radiation from these atoms is what is observed in AES.

The probability of transitions from a given energy level of a fixed atomic population was expressed by Einstein in the form of three coefficients, termed transition probabilities: $A_{ji}$, spontaneous emission; $B_{ij}$, spontaneous absorption; and $B_{ji}$, stimulated emission, which can be considered as representing the ratio of the number of atoms undergoing a transition to an upper level to the number of atoms in the initial or lower level and can be represented as follows:

$$N_j = N_0 \frac{g_j}{g_0} \exp(-\Delta E/KT) \tag{1.9}$$

where $N_0$ is the number of atoms in the lower state (or ground state, in most analytical work); $N_j$ is the number of atoms in the excited or upper level; $g_j$ and $g_0$ are the statistical weights of the $j$th (upper) state and 0, the ground state, respectively; $\Delta E$ is the difference in energy in Joules between these two states; $K$ is the Boltzmann constant ($1.38066 \times 10^{-23}$ J/K); and $T$ is the absolute temperature.

If self-absorption is neglected, then the intensity of emission, $I_{em}$, is

$$I_{em} = A_{ji} h \nu_{ji} N_j \tag{1.10}$$

where $h$ is Planck's constant ($6.624 \times 10^{-24}$ J s), and $\nu$ is the frequency of the transition ($\Delta E = h\nu$).

**Table 1.1**  Detection Limits for Diode Laser Flame/Graphite Furnace (ng/mL)

| Element | Flame | Graphite furnace |
|---------|-------|------------------|
| Li      | 3     | 5.7              |
| K       | 3     | 0.5              |
| Rb      | 10    | 0.3              |
| Cs      | 32    | 20               |

Therefore, $N$ is directly related to the concentration of the solutions as follows:

$$N_j = N \frac{g_j}{g_0} \exp(-\Delta E/KT) \tag{1.11}$$

The emission intensity of a spontaneous emission line, $I_{em}$, is related to this equation (sometimes called the Maxwell-Boltzman equation) as follows:

$$I_{em} = A_{ji} h \, \nu_{ji} N \frac{g_i}{g_0} \exp(-\Delta E/KT) \tag{1.12}$$

It can be seen that the atomic emission intensity is dependent on temperature and wavelength. Thus, a higher temperature at a longer wavelength would give the most intense atomic emission signal.

A plot of emission intensity against sample concentration will be linear. AES (and atomic fluorescence spectrometry) has linearity extending up to five to seven orders of magnitude compared with two to three orders of magnitude for AAS.

### 1.6.1 Instrumentation

The primary components of an atomic emission spectrometer are shown in Figure 1.9. While there is some similarity to the spectrometers used for AAS, optimum performance requires different components.

### 1.6.1.1 Excitation Sources

A large and varied number of excitation sources has been used in AES. It has been found that certain excitation sources may suit an application better than other excitation sources.

#### 1.6.1.1.a Flame

The flame was the traditional source in AES and was first used by Bunsen and Kirchoff in the 1860s. It was the major source until the development of the arc

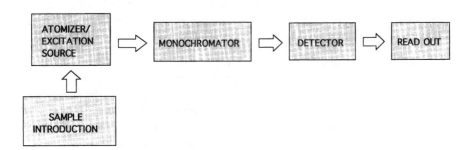

**Figure 1.9**   The primary components of an atomic emission spectrometer.

and spark in the 1940s. While still useful for specific applications such as for sodium and potassium in blood, it has been surpassed by plasma.

### 1.6.1.1.b Plasma

The development of the plasma excitation source for analytical AES has been a major advancement in atomic spectroscopy. Its higher temperature has made this source the choice for many applications.

***1.6.1.1.b.I Inductively Coupled Plasma.*** The most common plasma source is the inductively coupled plasma (ICP) which was first developed in the mid-1960s by Fassel and coworkers at Iowa State University, and Greenfield and coworkers at Albright and Wilson, Ltd. in England. It became commercially available in the mid-1970s. A typical ICP is shown in Figure 1.10 and consists of three concentric quartz tubes. These are frequently referred to as the outer, intermediate, and inner or carrier gas tubes. The outer tube can be various sizes, ranging from 9 to 27 mm. A two- or three-turn induction coil surrounds the top of the quartz tube or torch and is connected to a radiofrequency generator. The coil is water cooled. Argon, typically at a flow rate of 1–2 L/min, is introduced into the torch, and the radiofrequency field is operated at 4–50 MHz, most typically 27.12 MHz, and a forward power of 1–5 kW, typically 1.3 kW, is applied. An intense magnetic field develops around the coil, and a spark from a Tesla coil is used to produce "seed" electrons and ions in this region. This induced current flowing in a closed circular path results in great heating of the argon gas, and an avalanche of ions is produced. Temperatures in an ICP have been estimated to be 8000–10,000 K. The high temperature necessitates cooling which is achieved by applying argon to the outer tubes at flow rates as high as 17 L/min. The sample is introduced, usually as an aerosol, through the inner tube and is viewed at a distance of 5–20 mm above the coil. The advantages of the ICP include high temperature, long residence times, presence of none or few molecular species, optical thinness, and few ionization interferences.

The last decade has seen a tremendous amount of effort applied to evaluating and understanding the ICP. There have been numerous studies on its mechanisms and characterizing variations of the system. The reader is referred to a recent book edited by Montaser and Golightly[10] which describes the current status of the ICP. An interesting review of the analytical figures of merit for ICP-AES is available from Mermet and Poussel.[57]

***1.6.1.1.b.II Direct Current Plasma.*** A direct current plasma (DCP) jet is a flowing, gas-stabilized electrical discharge that is maintained by a core consisting of a continuous direct current arc. An electrical discharge is struck between a cathode and an anode. Tangentially flowing cooling gas enters the discharge chamber, causing a vortex flow around the cathode. The electrical conductivity of the gas is a function of the temperature. The cool, tangentially flowing gas in the chamber has a relatively low electrical conductivity, and the applied current preferentially flows as close to the chamber as possible. This is referred to

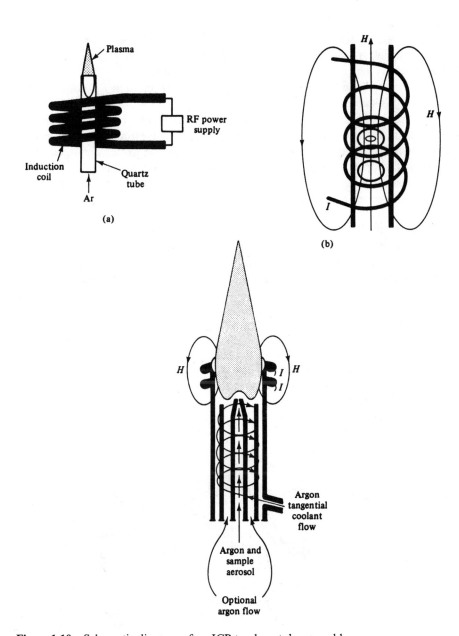

**Figure 1.10**   Schematic diagram of an ICP torch or tube assembly.

as the "thermal pinch" effect of the cooling gas, and it constrains the direct current arc core to exist in a narrow volume path. The construction of the direct current arc core is aided by the cylindrical orifice cathode out of which the discharge gas is expelled. The exhaust plume of the discharge constitutes the plasma, which is used as the excitation source. The thermal pinch effect results in an extremely high temperature in the arc core. Consequently, the temperature of the arc core and the plasma are significantly higher than for conventional arcs and gas-stabilized discharges.

The commercially available system is shown in Figure 1.11. It is an improved version consisting of the original system described above and a modified system that involves the use of two separate flowing gas electrodes arranged in an inverted V-shaped plasma, with sample introduction into the apex of the V. This system had problems with stability and drift, but these were solved by adding the carbon second electrode and placing the tungsten cathode vertically above the intersection of the two anode plasma columns as shown in figure 1.11. This configuration gives a well-defined and highly stabilized plasma column. The resulting excitation region has a low background luminosity and high temperature which make it well suited for spectrochemical analysis. It differs from other electrode-based discharges in the positioning of the electrode, the path of

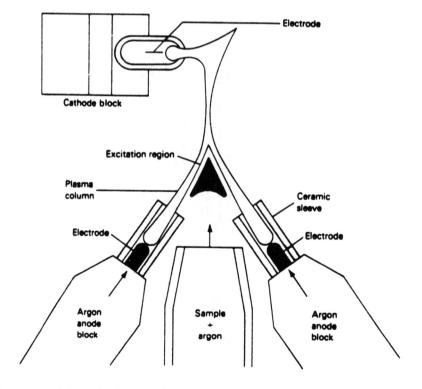

**Figure 1.11**   Schematic diagram of direct current plasma.

the flowing gas (which is a critical and fundamental feature of the plasma), and the location of the analytical observation region.

The DCP has been successfully applied to many analyses, achieving low detection limits, good precision, and accuracy. It is particularly stable in the presence of samples that contain a high proportion of dissolved solids, such as wear oils, clinical specimens, sea waters, and so forth. It is not as widely accepted or used as the ICP, perhaps due to the fact that it has only been available from one manufacturer, and the analytical performance characteristics, in general, are slightly inferior to the ICP. A fuller description of the DCP and its application is available from Sneddon.[58]

***1.6.1.1.b.III Microwave-Induced Plasma.*** The microwave-induced plasma (MIP) is the least utilized of the plasmas for atomic emission spectrometry. It has found use and potential as a detector for gas chromatography (GC). A schematic diagram of the MIP is shown in Figure 1.12. The plasma is initiated by providing seed electrons (as for the ICP) from the spark of a Tesla coil. The electrons oscillate in the microwave field and when sufficient energy is attained will ionize the support gas, usually helium. A microwave frequency of 2450 MHz is used. MIP is attractive because of its low power requirement (a few hundred Watts or less), and its less complex system (operated at atmospheric conditions) compared to other plasmas. The disadvantage is that it will not tolerate solvents and can be easily extinguished. The detection limits achieved by MIP-AES are, generally, inferior to the ICP-AES.

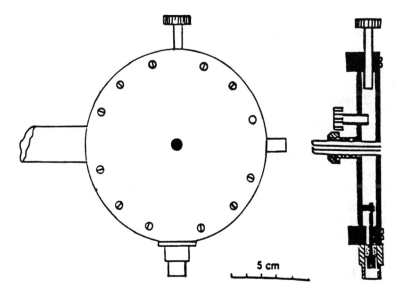

**Figure 1.12**   Schematic diagram of microwave-induced plasma.

***1.6.1.1.b.IV Other Plasma.***   The capacitively coupled plasma (CCP) is operated with argon or nitrogen as a support gas at flow rates around 5 L/min. Microwave powers of 200–100 W at 2450 MHz are required. The system is more tolerant of solvents.

### 1.6.1.1.c  Arc and Spark

Arc and spark sources have been popular in atomic emission spectrometry since the 1920s, with many developments occurring in the late 1930s through the late 1940s, and culminating in excellent commercial systems. Their advantage is in their ability to identify, primarily on a qualitative and semiquantitative basis, most elements in the periodic table directly in solids. This method has found widespread use and acceptance in the steel industry. The disadvantage is that it is generally not as sensitive or precise as conventional solution analysis, and works only for conducting materials. Several texts and articles are available on arc and spark AES.[59,60]

### 1.6.1.1.d  Other Excitation Sources

Various other excitation sources have been proposed including exploding thin films, the laser microprobe, and so forth. They are usually best suited to a particular need or analysis and have not been widely adopted. For the most part, they are not commercially available at present.

## *1.6.1.2 Sample Introduction*

Sample introduction devices are similar to the one discussed in chapter 1.4.3. A recent issue of *Spectrochimica Acta* (50B, numbers 4–7, June 1995) is devoted entirely to the subject of sample introduction in atomic spectrometry. It consists of 28 papers or technical notes by various authors or groups of workers involved in cutting edge research on sample introduction.

## *1.6.1.3 Wavelength Selection Devices*

The higher temperature of plasma will lead to a richer spectra with many more lines. In order to separate these lines and prevent or minimize spectral interferences, a high resolution monochromator is required. The 0.20-m grating typically used in AAS does not provide the resolution required. The most widely adopted system is the Echelle monochromator which uses high diffraction orders and large angles of diffraction. Resolution is around 0.015 nm compared to around 0.2 nm for a typical AAS monochromator. A polychromater is used for multi-element detection.

## *1.6.1.4 Detectors*

During the development of the plasma as an excitation source in analytical AES, PMTs were used. However, they have very limited use in multichannel systems, and AES quickly adopted other systems capable of multi-element analysis. In the early 1980s, the photodiode arrays (PDAs) and image sensing vacuum tubes, the so-called vidicons, were used. A PDA consists of arrays (256, 512, 1024, or 2048 elements arranged in a linear manner) of photodiodes operated on a charge transfer storage mode. Each diode is sequentially integrated (several microseconds) after all the diodes have been integrated with the incident radiation. The current generated by each photodiode is proportional to the intensity of the radiation it receives. The sequential measurement of the current can occur many times a second under the control of a microprocessor. This digitized information can be stored in a computer for electronic processing and visual display. Diode array systems are excellent for studying transient signals such as those of the laser-induced plasma with gate delay generator systems described in Chapter 5. However, they do have a somewhat limited resolution, usually 1–2 nm. Diode arrays are used in vidicons in the form of a spectrum. These are similar to a small television picture tube.

From the late 1980s through the present, there has been an interest in the use of the charge transfer device (CTD), specifically the charge coupled device (CCD) and to a lesser extent the charge injection device (CID). These are solid-state sensors that have integrated circuits. The charge generated by a photon is collected and stored in a capacitor. A typical pixel arrangement is $512 \times 320$ CCDs, although much larger arrangements, such as $2000 \times 2000$ pixels, have been constructed. The capacitor can be reverse biased by a positive voltage applied to the electrode, creating a potential well. The photons striking the array provide electron-hole pairs, and the electrons can be stored for a short time in the well. The amount of charge accumulated is a direct function of the incident radiation and time, and is very linear. The charge is shifted horizontally and down to a readout preamplifier which results in a scan of each row in series. CCDs are very useful where low levels of radiation are to be detected. At high levels, "blooming" occurs, which results in curvature of the response. A CID is a two-dimensional array of pixels. The photons generate positive charges below the negative well capacitors. Again, the amount of charge is proportional to the incident radiation. There has been rapid development of the CTD, and a more comprehensive text on the subject is available.[61]

## 1.7 Recent Developments in Atomic Emission Spectrometry

Considerable improvement and refinement in plasma source AES has occurred over the last decade. Improved detection limits have been achieved by rotating the plasma through 90 degrees, and the development of the miniature ICP. Considerable effort has been expended in the area of sample introduction (see earlier). Improved software has pushed ICP-AES to be a well-established and frequently used technique, particularly for multi-element AES. Since the early 1980s, the most significant improvement has been the development of plasma source mass spectroscopy, in particular the ICP-MS, and to a lesser extent, glow discharge mass spectroscopy (GD-MS). These developments are discussed under related techniques.

## 1.8  Atomic Fluorescence Spectrometry

Atomic fluorescence spectrometry (AFS) is based on the excitation of atoms by radiation of a suitable wavelength (absorption), and the detection and measurement of the resultant deexcitation (fluorescence). The only process of analytical importance is resonance fluorescence, in which the excitation and fluorescence lines have the same wavelength. Nonresonance transitions are not particularly analytically useful and involve absorption and fluorescence photons of different energies (wavelength).

The intensity of fluorescence, $I_f$, radiation between the ground state, $i$, and an excited state, $j$, is described by the following equation:

$$I_f = (I/4\pi) \; Y_{ji} I_{vij} \, k_v d_v \tag{1.13}$$

where $I$ is the distance between the excitation cell and the detector, $Y_{ji}$ is the quantum yield, $I_{vij}$ is the intensity of the excitation frequency $ij$, and $k_v \, d_v$ is the integrated absorption coefficient over the absorption line. The integrated absorption coefficient is proportional over the absorption to the concentration of the states $i$ and $j$, and to their statistical weights, the intensity of exciting photons, and the Einstein absorption coefficient, $B_{ij}$. Therefore, $I_f$ is a linear relationship between the intensity of the incident radiation and the quantum yield of fluorescence, as long as $I_{vij}$ is sufficiently below the level of saturation. In other words, $I_f$ is directly proportional to the intensity of the incident light. If the intensity of the incident light is increased, then the intensity of fluorescence will increase proportionally. A high intensity light source will result in a low detection limit. The problem in AFS has been to obtain a sufficiently stable, reproducible, and high-intensity light source that completely covers the useful electromagnetic spectrum (190–800 nm).

An interesting and informative review of AFS is available from Butcher.[62]

### 1.8.1  Instrumentation

The primary components of an atomic fluorescence spectrometer are shown in Figure 1.13. There are, as expected, many similarities between the AFS components and the components used for AAS and AES. The following discussion on AFS will highlight the instrumentation that is somewhat specific for AFS. It should be noted that the light source and detection system are at right angles to each other in AFS to minimize stray light.

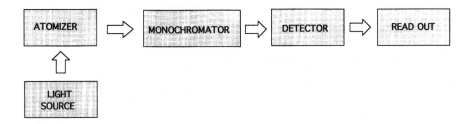

**Figure 1.13**   The primary components of an atomic fluorescence spectrometer.

## *1.8.1.1 Radiation Sources*

The primary objective of the radiation source in AFS is that of intensity, as opposed to AAS where the major objective of the radiation source is to emit a resonance radiation of the particular element, with a half-width less than the width of the absorption line; stability is a major concern. Clearly, the high intensity of the laser has shown great promise as a radiation source for AFS. This is discussed in detail in Chapter 3. Several radiation sources have been proposed for AFS, such as exploding wires, metal vapor lamps, and so forth, but these have never been widely used. The use of an ICP as an excitation source in conjunction with another atomizer such as a flame has shown some promise in AFS. The EDL is about the only conventional source currently used in AFS.

### 1.8.1.1.a Electrodeless Discharge Lamp

The electrodeless discharge lamp (EDL) was the most favored and developed source in AFS from the early 1960s until around 1980. This area has been reviewed by Sneddon et al.[63] The microwave-exited EDL was the most promising source for AFS, but problems of reproducibility and lack of commercial availability limited it uses, and since the early 1980s it has been little used. The radiofrequency EDL is discussed in the section on AAS. While commercial availability made the system more accessible, it was never adopted or widely accepted in AFS.

### 1.8.1.1.b Hollow Cathode Lamp

The HCL is described in the section of AAS. The low intensity output led to reduced or poor detection limits and limited its use in AFS. Various modifications of HCL's, often called boosted HCL's, showed some promise but was never widely used.

### 1.8.1.1.c Continuum Source

The attractiveness of the continuum source for AFS lay in its potential for multi-element AFS. This use showed some potential in the mid-1970s, however, some unreliability of the continuum source and the potential of lasers have limited its use in AFS.

### 1.8.1.2 Atomizer/Excitation Sources

Since its introduction in the early 1960s, analytical AFS was mostly used with the flame. The major difference with the flame used in AAS is that there is no need to have a long path length. Therefore, flames used for AFS have evolved into circular systems. Flames for AFS are best isolated from air (actually oxygen) to prevent quenching of the signal. This can be achieved with inert gas or even a physical presence, such as pyrex glass, around the flame.

The use of the electrothermal atomizer for AFS has shown improved performance compared to the flame.

## 1.9  The Practice of Atomic Spectroscopy

The choice of which analytical atomic spectroscopic technique to use will depend on the analysts' needs and expectations, and the sample. There are many and varied commercial systems available, but the analyst may decide that his or her needs are best filled by a laboratory-constructed system. The size of sample, whether it is a solid, liquid, or gas, the level to be detected, the accuracy and precision that is acceptable, the availability of a particular system, cost per sample, and the speed of analyses are factors to be considered.

Atomic spectroscopic methods are techniques that depend on the comparison of signals obtained from samples with those obtained from sample standards of known composition. In most cases these standards are aqueous solutions of the elements of interest. However, the analyses of real samples is complicated by the fact that the element of interest is present as part of a sample matrix. The matrix can cause an interference in the analysis. Therefore, in analytical atomic spectroscopy, much attention is paid to the possibility of interferences which can lead to reduced or poor accuracy. Accuracy can be defined as how close the atomic spectroscopic analysis is to the "correct" answer. In a typical method development, accuracy will be established via several or many ways including standard additions, comparison of the results of the atomic spectroscopic analyses with the results from a different method, or applying the atomic spectroscopic method to standard samples such as those supplied by the National Institutes of Science and Technology (NIST), Gaithersberg, Maryland. A concern of analytical atomic spectroscopy is precision, which can be defined as the repetitive analyses of a particular sample expressed as a percent. Precision will vary with many factors including sample, level to be determined, and choice of instrumentation. Finally, the detection limit is an important factor in analytical atomic spectroscopy. Current atomic spectroscopic techniques have detection limits in the microgram per liter to microgram per milliliter range. However, lower detection limits are possible with newer and improved techniques in analytical atomic spectroscopy. Several of these techniques using lasers are discussed in this book.

The reader is referred to a recent book by Butcher and Sneddon[64] which describes the practice of graphite furnace atomic absorption spectrometry. Much of the advice and suggestions in the book could be equally applied to many areas of analytical atomic spectroscopy.

## 1.10  Techniques Similar to Atomic Spectroscopy

As mentioned in section 1.7 two major developments in atomic spectroscopy has been plasma source mass spectroscopy and glow discharge mass spectroscopy. These techniques do not fall into the category of conventional atomic spectroscopy (i.e., AAS, AES, or AES), but they are closely linked and, frequently, their development has been achieved by atomic spectroscopists. This section also describes selection related techniques.

### 1.10.1  Inductively Coupled Plasma Mass Spectroscopy

Since the early to mid-1980s, the ICP has been used as the ion source for mass spectrometry when determining elements. Its advantages include from two to three orders of magnitude improvement in sensitivity compared to traditional ICP-AES; the mass spectra of the elements are very simple and unique, giving high specificity; inherent multi-element coverage; and the measurement of elemental isotopic ratios. Disadvantages include spectral interferences from molecular species and the increased cost and complexity of instrumentation.

An inductively coupled plasma mass spectroscopy (ICP-MS) system consists of the ion source, which is the ICP, an interface system, which consists of a sampling cone, a differentially pumped zone and a skimming zone, ion lenses, a quadrupole mass spectrometer, and a detector.

A detailed description of the instrumentation and the performance of ICP-MS is described elsewhere.[65,66]

### 1.10.2  Glow Discharge Mass Spectroscopy

Gas discharges used at reduced pressure have potential as an excitation source for AES, for direct solid sampling in AAS and AFS, and as a source for mass spectroscopy (MS). The glow discharge lamp is closely related to the hollow cathode lamp. The gas discharge or glow discharge is generated between two planar electrodes in a cylindrical glass tube. This tube is at a reduced or low pressure of a few torr of an inert gas. The sample of interest replaces the electrodes in the tube, and a complete spectra of the sample is produced, giving it multi-element capability. Disadvantages include that the sensitivity is about the same as conventional AES, and the sample turn-around can be quite time consuming (replacing sample, reducing the pressure, and introducing the inert gas). A detailed description of the glow discharge is available elsewhere.[67]

### 1.10.3. Resonance Ionization Spectroscopy

A typical resonance ionization spectroscopy (RIS) instrument consists of an atom source region, laser ionization region, and ion detection system. A block diagram of the instrument is shown in Figure 1.14. Atoms can be created in several ways including thermal, photodissociation of atoms from molecular vapors, making use of free atoms in a noble gas, laser vaporization, and ion or electron beam bombardment. The ionization region consists of the laser system producing the appropriate wavelength of light for the step transition to the ionization potential of the atom of interest. Once ionized, detection of ions can be accomplished by several means including proportional counters, Geiger-Mueller counters, electron multipliers, mass spectrometry (all types including magnetic sector, time-of-flight, quadrupole, and ion traps), and accelerator.

An accelerator experiment at Oak Ridge National Laboratory, Oak Ridge, Tennessee, was the beginning of RIS. While looking for a method to probe the metastable states of helium atoms, selective wavelengths of laser radiation were

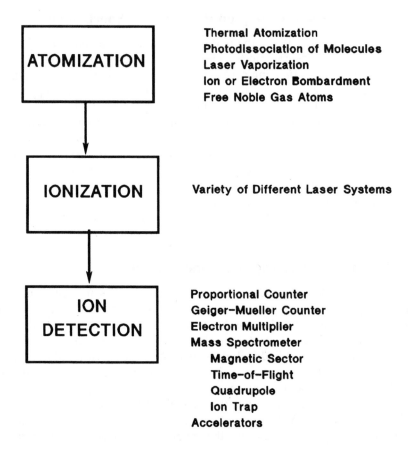

**Figure 1.14**   Block diagram of resonance ionization spectrometer.

used to promote the metastable electrons past the helium's ionization potential and thereby ionize the helium atoms. Once ionized, the helium atoms were readily detected using conventional ion detection systems. Following the helium experiment, RIS experiments on ground-state alkali atoms were undertaken because the wavelength required for the helium probe was very similar to that needed to ionize the alkali metal atoms. This work ultimately lead to the detection of a single Cs atom as it diffused away from a metal surface. This method was subsequently used in the detection of Cs atoms as a radiative decay product of Californium.[68] In addition to elemental detection, RIS has been applied to the measurement of cross section for photoionization of prepared quantum states of atoms, collisional line broadening in photoexcitation processes, diffusion of free atoms, the determination of chemical reaction rates for free atoms, the statistical behavior of free atoms, tests of the ergotic hypothesis for free atoms, photodissociation of diatomic molecules, fanofactor for electrons in gases, and analysis of selected impurities in materials such as semiconductors.[69]

Ionization of the free atoms once they are created can be accomplished by several methods, all of which involve laser systems. Direct ionization by laser energy is usually not practical except for alkali metal atoms whose ionization potentials are low. A more usual method of ionization is a stepwise method, as shown in Figure 1.15 which describes possible ionization pathways for germanium ionization methods can be divided into five major categories as shown in Figure 1.16. By the use of these five ionization schemes, it should be possible to detect 98 of the first 102 elements in the periodic table. This information is summarized in a review by Young et al.[70]

### 1.10.4. Resonance Ionization Mass Spectroscopy

Resonance ionization mass spectroscopy (RIMS), also referred to as resonantly enhanced multiphoton ionization spectroscopy (REMPI), is a method of ionization using laser light to excite electrons of a specific element first into higher electronic states, and eventually ionizing the atom. Because each element has a specific electronic structure, with a combination of laser energy, the element can be selectively ionized. This selectively ionized element is subsequently analyzed with a mass spectrometer in which additional selectivity, both elemental and isotopic, is added.

Resonance ionization techniques have been recognized since their introduction in 1977 as both selective and extremely sensitive, with demonstration of single atom detection for atoms in the region of laser excitation. The use of this technique is possible for all elements in the periodic table except He and Ne, whose electron configurations are not able to be excited by conventional laser systems. For some elements, namely H, C, N, O, F, P, S, Cl, Ar, As, Se, Br, Kr, I, Xe, At, and Rn, the first excited state that is accessible from the ground state with an optical transition is at a wavelength below 200 nm.[71] The selectivity of the technique is linked to the resonance ionization scheme employed for the laser ionization process. It should be noted that conventional RIMS does not

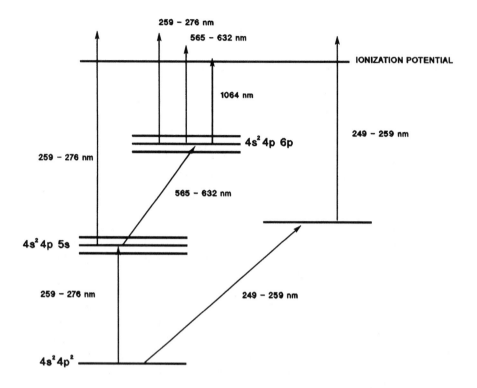

**Figure 1.15** Resonance ionization schemes for germanium.

give isotopic information, meaning that the allowed electronic transitions differ by such a small energy difference that RIMS is nonselective except in very special situations, and therefore, it is the mass spectrometer that determines isotopic ratios. However, since the ionization is selective, a high resolution mass spectrometer is not needed. The sensitivity of the technique is presently limited by the laser duty cycle and ion source brightness.[72]

There are three main types of atomizers used in RIMS, namely, thermal atomizers, sputter-initiated atomization, and laser atomization. The advantages of thermal vaporization are good stability and reproducibility, but the technique lacks spatial resolution and, therefore, gives no information on sample homogeneity. The other two methods are more efficient in the use of the sample and give good spatial resolution. Their main drawbacks are increased cost and complexity.

Thermal atomization can be achieved using a conventional electrothermal atomizer, such as a graphite furnace, or by placing the sample on a filament. While the duty cycle for pulsed operation of a filament is better than that of a graphite furnace, it is much lower than that of sputtering or laser atomization, and produces complex matrix effects requiring complex, element-specific separation schemes.

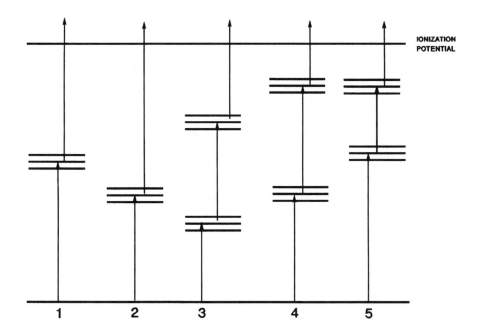

**Figure 1.16**    Five major resonance ionization schemes for ionization.

A second method of atomization is ion sputtering. Ion sputtering has several advantages over thermal atomization. First, complex samples are not subject to selective vaporization, therefore the analysis is representative of the actual sample. A second advantage of sputtered atomization is the flexibility in choice of substrate. Thermal atomization requires that the substrate be a refractory metal, whereas sputtering allows nonrefractory materials to be analyzed. The third possible advantage is the ability to pulse the sputtering source with the pulsed laser ionization source, thus overcoming the duty cycle mismatch that is suffered in thermal atomization, even with a pulsed filament. This fact should make sputtering vaporization two to five orders of magnitude more sensitive.

The third method of atomization is laser vaporization. Laser vaporization has seen increased interest due to its ability to perform direct solid analysis. To employ RIMS in conjunction with laser ablation requires two entirely separate laser systems fired with an adjustable time delay. A time-of-flight (TOF) mass spectrometer has been employed to study the laser ablation of Ta metal and the laser damage to CaF.

### 1.10.4.1  Application of RIMS for Multi-element Analysis

RIMS for multi-element analysis has been undertaken with limited success. To date, only thermal analysis has been attempted. Thermal atomization presents several obstacles to rapid analytical determinations. The first is that each element has an optimum vaporization temperature. The optimum temperature does not necessarily reduce the amount of background that may be present. If the element with the lowest optimum vaporization temperature is measured first, and that with the highest is measured last, rapid movement of the laser's wavelength is required because the matrix effects impact the optimum vaporization temperatures of the elements of interest. In addition to temperature and wavelength considerations, the mass spectrometer is also very important. If a quadrupole is used, single-ion monitoring could be used and the small area of the spectrum would make it possible to determine interelement interferences and allow background subtraction to be possible. Time-of-flight mass spectrometers are probably the best mass spectrometers for RIMS multi-element analysis because of their high transmission efficiency and capability for monitoring many different elements with each laser shot. The elements that can be probed using the frequency-doubled four major dyes are shown in Table 1.2.

### 1.10.5. Thermal Lens Effect Using Lasers

The thermal lens effect was accidently discovered in the mid-1960s when a Raman scattering cell filled with an organic dye was placed inside the cavity of a

**Table 1.2**  Atomic Transitions Available Using Specific Laser Dyes for RIMS (Frequency doubled)

| Element | R6G | RB | R101 | DCM | Element | R6G | RB | R101 | DCM |
|---------|-----|-----|------|-----|---------|-----|-----|------|-----|
| Al      |     |     | X    |     | Ni      | M   | X   | X    | X   |
| Ba      |     |     | X    |     | Os      | X   | M   | M    |     |
| Bi      |     |     | X    |     | Re      | X   | X   | X    |     |
| Co      | M   | M   | X    | X   | Rh      | M   | M   | X    | X   |
| Dy      | X   | X   | M    | X   | Ru      | X   | X   | M    | M   |
| Er      |     | X   |      |     | Sc      |     | X   | X    | X   |
| Eu      | X   | X   | X    | X   | Sr      |     | X   |      |     |
| Gd      | M   | M   | X    | X   | Ta      | X   | X   | X    | X   |
| Ga      | X   | M   |      |     | Th      | X   | X   | X    | X   |
| Hf      | X   | X   | X    | X   | Tm      | X   | X   | X    | X   |
| In      | X   | X   | X    | M   | Sn      | X   | M   | M    | M   |
| Fe      | X   | X   | X    |     | Ti      |     | X   |      | X   |
| Pb      | X   |     |      |     | W       | X   | M   | M    | M   |
| Li      |     |     |      | X   | U       | X   | X   | X    | X   |
| Lu      | X   | X   | X    | X   | V       | X   | X   | X    | M   |
| Mg      | X   |     |      |     | Y       | X   | X   | M    |     |
| Mn      | X   |     |      | X   | Zr      | X   | X   | M    | X   |
| Mo      |     | X   | X    | X   |         |     |     |      |     |

R6G—Rhodamine 6G, 278–290 nm
RB—Rhodamine B, 292–304 nm
R101—Rhodamine 101, 302–314 nm
DCM—310–325 nm
X—ground state originating
M—excited state originating

He-Ne laser in order to increase the scattering signal.[73] The researchers observed that the small amount of laser energy deposited in the liquid sample caused local heating. The heating, in turn, formed a refractive-index distribution that behaved as a diverging lens. Most liquids have a positive coefficient of thermal expansion and a negative temperature coefficient of the index of refraction, thus providing to the properties a diverging lens; the thermal lens defocussed the laser beam. Although the effect was originally thought to be a nuisance, thermal lensing has since been shown to be a valuable spectroscopic tool.[1] Examples of this include the ability to measure extremely small absorptions in seemingly transparent liquids. The method has also been used to increase the understanding of nonradiative processes in luminescence spectroscopy, vibrational-translational relaxation, thermal diffusion in gases, and photolytic reactions. In subsequent sections, the basic instrumentation and applications of this technique, as well as the theory behind the method, will be discussed in detail.

Initial experimentation done by Dovichi and Harris[74] set the basis for the theory behind the laser-induced thermal lens effect. The thermal lens effect produced in their experimental arrangement was similar to that used in normal absorption spectroscopy. The major difference was that laser radiation passing through the sample was detected only at the center of the beam by restricting the field of view of the detector with a pinhole. If the path of the laser being used is initially blocked and opened with a shutter, the thermal lens effect takes a finite period of time to build up. A steady-state condition is reached when the rate of laser heating of the liquid sample equals the rate of heat loss due to the thermal conductivity of the solvent being used. The buildup of the lens effect takes tens of microseconds to hundreds of milliseconds, depending on the thermal conductivity of the solvent and the radius of the laser beam traversing the sample.

The intensity of the radiation at the beam center $I(t)$, will initially (at $t = 0$) reflect only the Beer's law response of the sample. That is:

$$I_{(0)}/I_0 = 10^{-A} \tag{1.14}$$

For the very small $A$ values that are normally encountered with this method, we obtain the formula:

$$[I_0 - I_{(0)}]/I_0 = 2.303\ A \tag{1.15}$$

where $A$ is the sample absorbance and $I_0$ is the incident light intensity. When the steady-state temperature is reached, the intensity at the detector, $I_d$, is very dependent on the optical arrangement of the system, but reaches a minimum value when the sample is placed one confocal length beyond the beam waist. When this is done, the following equation governs the initial and final intensities:

$$(I_{(0)} - I_d)/I_d = -2.303P\ (dn/dT)A/\lambda k \tag{1.16}$$

$$= 2.303\ EA \tag{1.17}$$

where $P$ is the laser power in watts, $dn/dT$ is the change in solvent refractive index with temperature, $\lambda$ is the laser wavelength, $k$ is the thermal conductivity of the solvent in watts per centimeter kelvin, and $E$ is the enhancement of this effect relative to Beer's law behavior. This expression assumes that all of the absorbed light is converted to heat.

Several key factors are shown from the above equations. First, the relative change in intensity due to the thermal lens effect depends linearly on laser power, similar to fluorescence. The method does, however, reach a saturation point with high laser fluxes, just as with fluorescence. Second, if the laser is pulsed, or chopped in the case of continuous wave lasers, a nearly simultaneous measurement of blank signal can be made in the presence of the sample. Although this is not a true blank because the intensity is reduced from $I_0$ by $10^{-4}$, this factor cancels in determining $(I(0) - I_d)/I_d$ since $I_d$ is reduced by the same proportional amount. Third, since the enhancement, $E$, is related to the thermoconductivity of the solvent, the choice of solvent governs the enhancement that is realized for a particular laser intensity and wavelength. A general trend is that the enhancement in more polar solvents suffers from the larger thermal conductivity and a small $dn/dT$. Fourth, because most of the initial work was done using CW He-Ne laser systems, the lack of tunability precluded the measurement of spectra; therefore, the selectivity of a method must be provided by chemical means with either a chromatogenic reaction or separation technique.

From this initial study, the field has expanded to include pulsed laser systems, dual laser systems, and systems using gaseous samples. With this expansion, theory has also expanded with the addition of more complex mathematical expressions for explaining the phenomena. For example, when a gaseous sample is used in conjunction with a pulsed laser system, the transient decay of the thermal lens signal, $S_p$, is given by:

$$S_p = S_{p(t=0)} (1 + 2t/t_c)^{-2} \tag{1.18}$$

$$S_{p(t=0)} = 2.303 \, E_p A \tag{1.19}$$

$$E_p = -3^{3/2} E_t (dn/dt)/\lambda_p (\omega_{op})^2 \, C_p \sigma \tag{1.20}$$

where $S_{p(t=0)}$ is the signal intensity at $t = 0$ after excitation, $t$ is the time, $t_c[(\omega_{1p})^2/4D]$ is the characteristic time constant, where $\omega_{1p}$ is the beam radius of the heating laser at the sample and $D$ is the thermal diffusivity. $E_t$ is the pulsed energy of the heating laser, $\lambda_p$ is the wavelength, $\omega_{op}$ is the beam radius at the waist, $\sigma$ is density, $C_p$ is specific heat, $(dn/dT)$ is the variation of refractive index with temperature, and $A$ is the absorbance of sample.[75]

In comparing a pulsed laser system to that of CW, we find that enhancement factors for normal organic solvents are 300–11,000 using 1 W of the CW laser source, and 200–600 using 1 mJ of a pulsed laser source.[76,77] It has been shown that the use of the pulsed laser is advantageous if the average power is identical to that of the CW laser at 1 Hz. But since the CW laser generally has a higher average power, the CW laser provides much more sensitive detection for most applications. On the other hand, the pulsed thermal lens system has very large enhancement factors and is quite useful for sensitive determinations of trace gaseous samples. It is noted that even if the average power of the pulsed laser is less than $10^{-3}$ times at 1 Hz in comparison with the CW laser, the enhancement factor is 10 times larger for the samples in air.

The reason that there is a large enhancement factor with a pulsed laser system is that a gaseous sample has a very low density and has a small heat capacity per unit volume. Therefore, the temperature increase is much larger in comparison with a condensed phase sample. The relatively small enhancement factor for CW excitation for a gaseous sample is due to a small variation in refractive index with temperature. Furthermore, the thermal diffusivity of air is larger than that of liquids because the air is less dense, and the induced thermal lens effect disappears rapidly.

Sensitivity of the thermal lens technique can be defined as a relative amplitude of the signal to that of conventional spectrophotometry, which is proportional to the enhancement factor explained above. The detection limit may be improved by increasing the enhancement factor; therefore, a heating laser with a large pulse energy is essential for trace analysis. The detection limit is also determined by the precision of the observed value. A low-repetition rate pulsed laser with a large pulse energy gives a large enhancement factor but gives relatively poor precision. CW lasers, on the other hand, give small enhancement factors, but repetitive signal acquisition can improve precision.[78]

The basic components of a modern thermal lens experiment consist of two lasers: one high-power laser to achieve rapid heating, for example, a nitrogen-laser pumped dye laser, and a He-Ne laser to be used as a probe beam. The collimated beams of both lasers are focused into the sample cell which can contain a gas or liquid sample. The laser beam that passes through the sample cell is focused onto a pinhole in front of a photomultiplier tube. Output from the photomultiplier is processed through a series of electronics and a microcomputer.

### 1.10.5.1 Applications of Thermal Lensing Spectroscopy to the Detection of Phosphorus

Methods for detection of phosphorus in environmental samples at nanogram levels are very limited. The most sensitive method is the classic molybdenum blue colorimetry based on the reduction of molybdenum in phospho-12-molybdate. It is capable of subnanogram level detection in pure aqueous standards. Testing of environmental samples, however, has not achieved this level of detection. Among the various attempts to improve molybdenum blue colorimetry is the use of thermal lensing colorimetry with an $Ar^+$ laser pumped dye laser. In previous studies of thermal lensing colorimetry, the detection limit of nitrite ions was improved about 500 times over ordinary colorimetry.[79] The detection limits for phosphorus (twice the standard deviation of five repetitions of color developments for blank solutions) were as follows: water, 72.7 pg/mL; 1:2 acetone-water mixture, 5.0 pg/mL; and 1:1 acetone-water mixture, 13.9 pg/mL.

For environmental samples such as seawater, precision phosphorus determination can be seen from results of 495 $\pm$ 10 pg/mL for a standard solution containing 500 pg/mL.

# References

1. Ingle, J.D. Jr.; Crouch, S.R. *Spectrochemical Analysis;* Prentice Hall: Englewood Cliffs, NJ, 1988.

2. Robinson, J.W. *Atomic Spectroscopy;* 2nd ed.; Marcel Dekker: New York, 1996.

3. Lajunen, L.H.J. *Spectrochemical Analysis by Atomic Absorption and Emission;* Royal Society of Chemistry: Cambridge, UK, 1992.

4. Willard, H.H.; Merritt, L.L. Jr.; Dean, J.A.; Settle, F.A. Jr. *Instrumental Methods of Analysis;* 7th ed.; Wadsworth: Belmont, CA, 1988; Chapters 5, 9, and 10.

5. Kennedy, J.H. *Analytical Chemistry: Principles;* 2nd ed.; W.B. Saunders: New York, 1990; Chapters 11 and 13.

6. Skoog, D.A.; West, D.M.; Holler, F.J. *Fundamentals of Analytical Chemistry;* 7th ed.; W.B. Saunders: New York, 1996; Chapters 22, 23, and 26.

7. Hassan, S.S.M. *Organic Analysis Using Atomic Absorption Spectrometry;* Ellis Horwood: Chichester, UK, 1984.

8. *Sample Introduction in Atomic Spectroscopy;* Sneddon, J., Ed.; Elsevier Science: Amsterdam, 1990.

9. *Atomic Absorption: Theory, Design, and Applications;* Haswell, S.J., Ed.; Elsevier Science: Amsterdam, 1991.

10. Montaser, A.; Golightly, D.W. *Inductively Coupled Plasmas in Analytical Atomic Spectrometry;* 2nd ed.; VCH: New York, 1992.

11. *Spectrochimica Acta Reviews* (formerly *Progress in Analytical Spectroscopy*); Pergammon: Oxford, England.

12. *Atomic Spectroscopy;* Perkin-Elmer: Norwalk, CT.

13. *The Spectroscopist;* Thermo Jarrell Ash-Baird: Franklin, MA.

14. *Journal of Analytical Atomic Spectrometry;* Royal Chemical Society: London.

15. *Advances in Atomic Spectroscopy;* Sneddon, J. Ed.; JAI: Greenwich, CT, 1992, Vol. 1; 1995, Vol. 2; 1997, Vol. 3.

16. Wollaston, W.H. *Phil. Trans.* **1802,** 92, 365.

17. Fraunhofer, J. *Gilbert's Ann.* **1817,** 56, 264.

18. Kirchoff, G.R. *Monatsber. Akad. Wissensch. Berlin* **1859,** October, 662.

19. Kirchoff, G.R.; Bunsen, R. *Pogg. Ann.* **1860,** 110, 161.

20. Wood, R.W. *Phil. Mag.* **1905,** 10, 513.

21. Walsh, A. *Spectrochimica Acta* **1955,** 7, 108.

22. Alkemade, C.T.J.; Melatz, J.M.W. *Appl. Sci. Res.* **1955,** B4, 289.

23. Alkemade, C.T.J. In *Proceedings of the Xth Coloquim Spectroscopicum;* Lippincott, E.R.,; Margoshes, M., Eds.; Spartan Books: Washington DC, 1963.

24. Winefordner, J.D.; Vickers, T.J. *Anal. Chem.* **1964,** 36, 161.

25. Winefordner, J.D.; Staab, R.A. *Anal. Chem.* **1964,** 36, 165.

26. West, T.S. *Analyst* **1966,** 91, 69.

27. Herzberg, G. *Atomic Spectra and Atomic Structure;* Dover: New York, 1945.

28. Sobelman, I.I. *Atomic Spectra and Radiative Transitions;* 2nd ed.; Springer-Verlag: Berlin, Heidelberg, and New York, 1992.

29. Harnley, J.M.; O'Haver, T.C.; Golden B.; Wolf, W.R. *Anal. Chem.* **1979,** 52, 2007.

30. Harnley, J.M. *Anal. Chem.* **1986,** 58, 933A.

31. King, A.S. *Astrophys. J.* **1905,** 21, 236.

32. L'vov, B.V. *Ing. Fiz. Zhur.* **1959,** 11, 44.

33. L'vov, B.V. *Spectrochim. Acta* **1961,** 17, 761.

34. Massmann, H. *Spectrochim. Acta* **1968,** 23B, 215.

35. Woodriff, R.W. *Appl. Spectrosc.* **1974,** 28, 413.

36. Williams, X.K.; West, T.S. *Anal. Chim. Acta* **1969,** 45, 27.

37. Anderson, R.G.; Maines I.S.; West, T.S. *Anal. Chim. Acta* **1970,** 51, 335.

38. Wirz, P.; Gross, M.; Ganz, S.; Scharmann, A. *Spectrochim. Acta* **1983,** 38B, 1217.

39. Ng, K.C.; Jensen, R.; Brechman, M.J.; Santos, W.C. *Anal. Chem.* **1988,** 60, 2821.

40. Lui, H.; Ren, Y.; Zhang, H.; Jin, Q. *Microchem. J.* **1991,** 44, 86.

41. Liang, D.C.; Blades, M.W. *Anal. Chem.* **1988,** 60, 27.

42. Browner, R.F.; Boorn, A.W. *Anal. Chem.* **1984,** 56, 786A.

43. Browner, R.F.; Boorn, A.W. *Anal. Chem.* **1984,** 56, 875A.

44. Olsen, K.W.; Haas, W.J. Jr.; Fassel, V.A. *Anal. Chem.* **1977,** 49, 632.

45. Wendt, R.H.; Fassel, V.A. *Anal. Chem.* **1968,** 38, 337.

46. Fry R.C.; Denton, M.B. *Appl. Spectrosc.* **1979,** 33, 393.

47. Heine, D.R.; Denton, M.B.; Schlaback, T.D. *Anal. Chem.* **1982,** 54, 81.

48. Sneddon, J. *Am. Lab.* **1991,** 23(6), 15.

49. Newstead, R.A.; Price, W.J.; Whiteside, P.J. *Prog. Anal. Atom. Spectrosc.* **1978,** 1, 267.

50. Sneddon, J. *Spectroscopy* **1987,** 2(5), 38.

51. Farah K.S.; Sneddon, J. *Appl. Spectrosc. Rev.* **1995,** 30(4), 351.

52. Kuhl, J.; Marowsky, G.; Torge, R. *Anal. Chem.* **1972,** 44(2), 375.

53. Wieman, C.E.; Hollberg, L. *Rev. Sci. Instrum.* **1991,** 62(1), 1.

54. Barber, T.E.; Walters, P.E.; Wensing, M.W.; Winefordner, J.D. *Spectrochim. Acta* **1991,** 46B, 1009.

55. Hergenroder, R.; Niemax, K. *Spectrochim. Acta* **1988,** 43B, 1443.

56. Groll, H.; Niemax, K. *Spectrochim. Acta* **1993,** 48B, 633.

57. Mermet, J.M.; Poussel, E. *Appl. Spectrosc.* **1995,** 49(10), 12A.

58. Sneddon, J. *Spectroscopy* **1989,** 4(3), 26.

59. Walters, J.P. *Appl. Spectrosc.* **1972,** 26, 323.

60. Walters, J.P. *Appl. Spectrosc.* **1972,** 26, 17.

61. *Charge Transfer Devices;* Denton, M.B. et al., Eds.; VCH: New York, 1993.

62. Butcher, D.J. *Spectroscopy* **1993,** 8(2), 14.

63. Sneddon, J.; Browner, R.F.; Keliher, P.N.; Winefordner, J.D.; Butcher, D.J.; Michel, R.G. *Prog. Anal. Spectrosc.* **1989,** 12(4), 369.

64. Butcher D.J.; Sneddon, J. *Practical Guide to Graphite Furnace Atomic Absorption Spectrometry;* John Wiley & Sons: New York, 1997.

65. Houk, R.S. *Acc. Chem. Res.* **1994,** 27, 333.

66. Beres, S.; Thomas, R.; Denoyer, E.; Bruckner, P. *Spectroscopy* **1994,** 9(1), 20.

67. Jabubowski, N.; Steuwer, D.; Vieth, V. *Anal. Chem.* **1987,** 59, 1825.

68. Hurst, G.S.; Nayfeh, M.H.; Young, J.P. *Appl. Phys. Lett.* **1977,** 30(5), 229.

69. Hurst, G. *Anal. Chem.* **1981,** 53, 1448A.

70. Young, J.P.; Hurst, G.S.; Kramer, S.D.; Payne, M.G. *Anal. Chem.* **1979,** 51, 1050A.

71. Sjostrom, S.; Maunchien, P. *Spectrochim. Acta Rev.* **1993,** 15, 153.

72. Kimock, F.M.; Baxter, J.P.; Pappas, D.L.; Kobrin, P.H.; Winograd, N. *Anal. Chem.* **1984,** 56, 2782.

73. Gordon, J.; Leite, R.; Moore, R.; Porto, S.; Whinnery, J. *J. Appl. Phys.* **1965,** 36, 3.

74. Dovichi, N.; Harris, J. *Anal. Chem.* **1979,** 51, 726.

75. Mori, K.; Imasaka, T.; Ishibashi, N. *Anal. Chem.* **1983,** 55, 1075.

76. Imasaka, T.; Miyaishi, K.; Ishibashi, N. *Anal. Chem.* **1980,** 115, 407.

77. Mori, K.; Imasaka, T.; Ishibashi, N. *Anal. Chem.* **1982,** 54, 2034.

78. Mori, K.; Imasaka, T.; Ishibashi, N. *Anal. Chem.* **1983,** 55, 1075.

79. Fujiwara, K.; Lei, W.; Uchiki, H.; Shimokoshi, F.; Fuwa, K.; Kobayashi, T. *Anal. Chem.* **1982,** 54, 2026.

# 2

# Lasers

*Joseph Sneddon, Yong-Ill Lee,*
*Xiandeng Hou, Jack X. Zhou,*
*and Robert G. Michel*

## 2.1 Introduction

For over 35 years after it was invented,[1,2] the laser has been involved in many aspects of analytical chemistry,[3-9] especially in the field of atomic spectroscopy, as already discussed in Chapter 1. For atomic spectroscopy, an ideal laser should be stable, tunable, noise free, reliable, inexpensive, lightweight, rugged and portable, easy to operate, and possess high output power. However, a real laser that meets all these requirements simultaneously does not exist. In practice, one has to specify those laser characteristics that are most important for one's experiment and then select a laser that best fits those characteristics.

This chapter is written and organized with lasers for atomic spectroscopy in mind. It describes the basic principles of lasers, the kinds of lasers that are available, and the characteristics that are most important from the atomic spectroscopists' standpoint. A nonmathematical approach is taken in this chapter. An additional aim is to provide an updated description of new types of lasers for those who use or plan to use lasers as spectroscopic and analytical tools.

## 2.2 Fundamentals of Lasers

### 2.2.1 Basic Concepts

The word *laser* is an acronym for *light amplification by stimulated emission of radiation.* In order to appreciate the concepts of lasing action, three fundamental phenomena are relevant. They are the processes of absorption, spontaneous emission, and stimulated emission.

### 2.2.1.1 Absorption

Consider an atom that has only two energy levels, an upper level $E_1$ and a lower level $E_0$, as shown in Figure 2.1, and assume that the atom is initially lying in the lower level. If $E_0$ is the ground level, the atom will remain in this level unless some external stimulus is applied to it. When the atom is irradiated by incident radiation with photon energy equal to the energy difference of the two levels, $E_1 - E_0$, there is a finite probability that the atom will be excited to the upper level, $E_1$. This is the absorption process.[10]

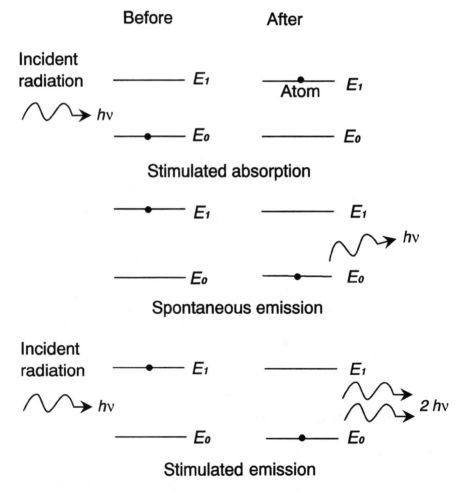

**Figure 2.1** Schematic illustration of various absorption and emission processes.

### 2.2.1.2 Spontaneous and Stimulated Emission

The atoms in the higher energy level, $E_1$, release their excess energy by either nonradiative or radiative processes, and return to the lower energy level $E_0$. Emission of photons can occur by two distinct processes: spontaneous or stimulated emission. Spontaneous emission occurs when the excited atom returns to the ground state by releasing the corresponding energy difference ($E_1 - E_0$) without any external stimulus.[10] The frequency $v$ of the radiated wave is then given by the expression

$$v = (E_1 - E_0)/h \qquad (2.1)$$

where, $h$ is Planck's constant. Light emitted in this way radiates from the atoms in random directions and at well-defined wavelengths. Stimulated emission occurs when additional photons that have the same energy as the absorbed photon trigger atoms in the excited state to emit radiation. Such stimulated emission has two important properties: It has the same energy and wavelength, as the stimulating photons, and it is in phase, or coherent, with the original light. These stimulated photons add to the incident wave constructively to produce the laser beam.

## 2.2.2 Principles of Lasing Action

A laser is an optical transducer that converts energy supplied by a pump source into ordered and intense light. All lasers consist of three main parts: an energy source for pumping, a lasing medium, and an optical cavity or resonator.

### 2.2.2.1 Pumping and Population Inversion

Consider two arbitrary energy levels of a given atom and let $N_1$, the lower state, and $N_2$, the higher state, be their respective populations. The ratio of populations is given by the Boltzmann relationship for a system in thermal equilibrium

$$N_2/N_1 = (g_2/g_1) \exp[(E_1 - E_2)/kT] \qquad (2.2)$$

where $k$ is Boltzmann's constant, and $T$ is the absolute temperature of the material. Under thermodynamic equilibrium, there is usually a larger population in a lower state than in a higher state, that is, $N_1 > N_2$. If a nonequilibrium condition is achieved such that $N_1 < N_2$, then the material will act as a light amplifier. This case is referred to as a population inversion, and the material is called an active material.[10] Hence, the minimum requirement for lasing action is population inversion (see Figure 2.2). The buildup of a population of atoms or molecules in an excited state within the medium can be achieved by an external energy source. This is known as pumping. There are two basic pumping mechanisms for lasers: optical and electrical. The pump method varies, although generally, gas lasers use electrical discharge, solid-state lasers such as Nd:YAG and ruby lasers use flash lamps, and liquid lasers such as dye lasers use other

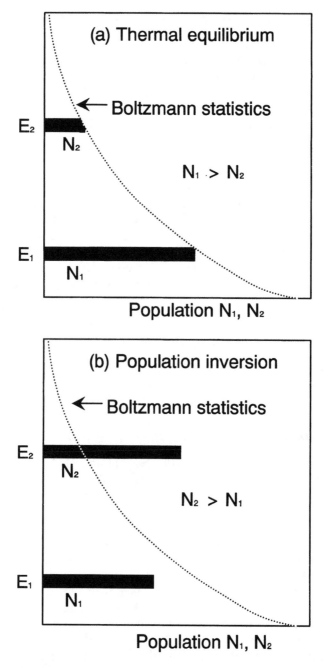

**Figure 2.2** Population profile at thermal equilibrium (*a*) and population inversion (*b*) by prediction with Boltzmann statistics.

lasers as the pump source. Arc lamps and tungsten lamps are used in many continuous lasers, while flash lamps are used in many pulsed lasers. If population inversion is achieved at a high enough level to overcome energy losses inside the cavity, threshold conditions will be met within the cavity and lasing will occur.

### 2.2.2.2 Resonators and Amplification

The optical resonator is a specially formed laser material (crystal or rod for solid lasers, or long tube or cell for liquid or gas lasers) that is closed off with two mirrors, one of which is semitransparent. Once a few atoms or molecules have emitted light by spontaneous emission, these emitted photons can stimulate emission from other excited atoms or molecules. A single passage of photons through the medium is not generally sufficient for stimulated emission to play a significant role. If stimulated emission occurs in the axis between the two mirrors, it is reflected back and forth through the tube to stimulate emission repeatedly from the medium. In this condition the light is said to be resonant with the cavity. At resonant wavelengths, the reflectivity inside the cavity can be very high, which creates a feedback environment. Photons spontaneously emitted in other directions are lost from the laser medium and no longer contribute to stimulated emission. In short, the lasing action develops by amplification of the spontaneous emission by stimulated emission.

Control of the laser is achieved by adjustment of the losses experienced by the resonator cavity. A buildup of energy occurs at the desired wavelengths and resonant frequency, while energy is lost at other wavelengths. There are several designs that use one or two concave spherical mirrors that simplify alignment compared to plane mirrors.

Laser resonators have two distinct types of modes: transverse and longitudinal. Transverse modes manifest themselves in the cross-sectional profile of the beam, while longitudinal modes correspond to different resonances along the length of the laser cavity that occur at different frequencies or wavelengths within the gain bandwidth of the laser. The transverse modes are characterized by two integers, $q$ and $r$, that are designated by $TEM_{qr}$, where $q$ is the number of minima or phase reversals as the beam is scanned horizontally, and $r$ is the number of minima as it is scanned vertically.[11] The simplest is the $TEM_{00}$ mode which exhibits a Gaussian profile where the intensity peaks at the center. The next higher-order mode is $TEM_{01}$, and so on. Some lower-order transverse patterns of a laser are shown in Figure 2.3.

### 2.2.3 Properties of Laser Beams

The important properties of lasers from the atomic spectroscopists' point of view are tunability, power density, monochromaticity, directionality, pulse-to-pulse stability for pulsed lasers, ease of use, and coherence.

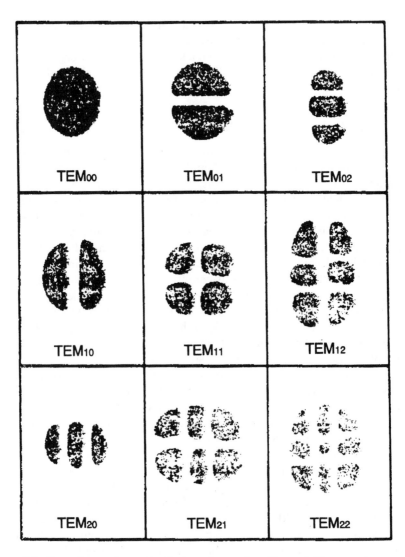

**Figure 2.3**  Some low-order modes that can be produced by a stable resonator.

### 2.2.3.1 Tunability

The wavelength tunability of lasers is very important in laser atomic spectroscopy. Three different tuning methods are available for tunable lasers. The first method, as used in dye lasers, involves the use of an active laser medium that gives sufficiently high gain over a wide spectrum. Dispersive elements are inserted inside the dye laser resonator to narrow the spectral output of the laser and to allow tunability to the proper wavelength. The second method, as used in diode lasers, utilizes a shift of the energy levels of the active medium by external fields or other interactions. The third method is based on optical frequency mixing techniques in a nonlinear medium to obtain tunable light sources, for example, the optical parametric oscillator (OPO) laser. In the first two methods, the lasers derive their gain from the stimulated emission generated by atomic and molecular transitions. Consequently, the inherent bandwidth of these transitions constrains the maximum tuning range of the tunable laser. However, in the last method, the laser output is derived from a nonlinear frequency conversion process rather than by stimulated emission, and the tunable range is determined by factors such as the transmission characteristics of the nonlinear material. Typical tunable ranges for dye lasers and OPO lasers are 340 nm to 1μm and 410 nm to 2μm, respectively.

### 2.2.3.2 Power Density

The power density of a laser of even moderate power, for example, a few milliwatts, is orders of magnitude greater than that of the brightest conventional sources. A pulsed laser produces $10^{15}$–$10^{30}$ photons/s, dependent on the average power of the laser, while a hollow cathode lamp may produce $10^6$–$10^8$ photons/s at a single wavelength. This can significantly enhance the signal-to-noise ratio (S/N), which is very important in analytical spectrometry for reduction of noise caused by the detector or background radiation. Typical powers of commercial continuous-wave lasers range from less than a milliwatt to tens of kilowatts, with comparable average powers available from pulsed lasers. Laser pulses with a peak power in excess of a terrawatt are now available from relatively small systems.[12] High power density allows a measurable amount of light to get through an optically dense sample so that high absorbances can be measured. The high intensity also allows for nonlinear spectroscopy such as saturation spectroscopy and multiphoton spectroscopy, and surface-ablation sampling for absorption, emission, photoionization, and mass spectroscopy.

### 2.2.3.3 Monochromaticity

Monochromaticity results because all the photons are emitted from a transition between the same two atomic or molecular energy levels and, hence, have almost exactly the same frequency. Nonetheless, there are considerable variations in linewidth of the emitted light. In some cases, the transition involves simultaneous changes in two or more quantum states, which leads to a range of wavelengths in the gain bandwidth of the laser. This situation is illustrated in Figure 2.4. One well-defined wavelength can be selected by an etalon, which is an optical element placed within the laser cavity in order to achieve the optimum monochromaticity. It is relatively easy to obtain an emission linewidth as low as 1 cm$^{-1}$ with continuous-wave lasers, and in frequency-stabilized lasers the linewidth may be four or five orders of magnitude smaller.

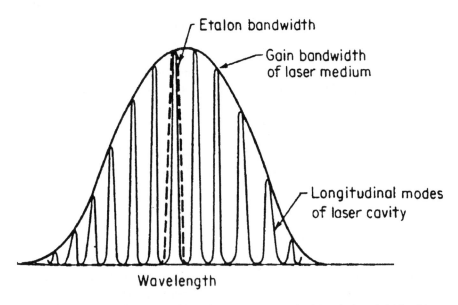

**Figure 2.4** Transmission bandwidth of etalon compared with gain bandwidth of laser medium and the longitudinal modes of a laser cavity.

### 2.2.3.4 Directionality

One of the striking features of most lasers is that the output is in the form of an almost parallel beam. This is a very useful feature for a number of applications since it means that it is straightforward to collect the emitted radiation and focus it onto a small area by use of fairly simple optics. Most laser beams are well collimated with no more than milliradian divergence, and most of the intensity occurs within a millimeter-scale cross section.

### 2.2.3.5 Pulsing

Pulse repetition rates vary widely. Ruby and neodymium-glass lasers can generate at most a few tens of pulses per second (pps), while excimer lasers can produce up to a thousand pulses per second, and copper vapor lasers several thousand pulses per second. Pulse repetition rate, duration, and peak power can be adjusted by a variety of laser accessories such as Q-switches, mode lockers, and cavity dumpers, described in Section 2.4. The temporal width of a pulsed laser is of importance to analytical chemists. Nanosecond ($10^{-9}$ s), picosecond ($10^{-12}$ s), and femtosecond ($10^{-15}$ s) pulses are commonly used to monitor very fast analytical processes. Details on the pulse characteristics of lasers are described in Section 2.3.

### 2.2.3.6 Coherence

Coherence is the property that most clearly distinguishes a laser beam from a conventional light source. There are two different kinds of coherence: spatial and temporal. Spatial coherence imbues the beam with excellent directionality and focusability, whereas temporal coherence provides excellent monochromaticity with narrow linewidths. Temporal coherence means that there is a fixed phase relationship between parts of the wave emitted at different times. If the phase changes uniformly with time, then the beam is said to have temporal coherence, although perfect temporal coherence is forbidden by the uncertainty principle. Also, laser radiation has spatial coherence, which means that a surface exists over which the phase of the wave is correlated at all points. These ideas are illustrated in Figure 2.5.

The useful quantities that quantify coherence are the coherence time, $t_c$, and the coherence length, $L_c$. These quantities vary with different types of laser, depending on the frequency bandwidth of the emission. The coherence length, $L_c$, equals the speed of light, $c$, divided by the frequency bandwidth of the laser, $\Delta v$

$$L_c = c/\Delta v \tag{2.3}$$

Also, the coherence time is related to the linewidth of the emission ($\Delta v$)

$$t_c = 1/\Delta v \tag{2.4}$$

## 2.3 Practical Lasers

Several criteria can be used in classification, but lasers are generally classified by laser material and pump method, such that lasers are grouped as dye, gas, semiconductor, and solid-state lasers.[13]

Dye lasers are usually excited by lamps or pulsed lasers for pulsed operation, or by continuous lasers for continuous wave (CW) radiation. Gas lasers are usually excited by a high-voltage glow discharge. Semiconductor lasers can be excited optically or by electron beams, or by passage of electrical current through a p-n junction. Solid state lasers are usually excited optically by pulsed discharge lamps or lasers for high-power pulses, and by continuous lamps or lasers for CW operation.

### 2.3.1 Dye Lasers

The operation of the liquid-state dye laser is based on fluorescent transitions in large organic molecules in solution, and its output wavelength can be tuned from the near-ultraviolet through the near-infrared range, typically 340 nm to 1 μm, depending on the dye used. The use of frequency-doubling crystals can extend emission further into the ultraviolet. Until recently, the most important lasers from an analytical point of view have been the dye lasers, but tunable solid-state lasers are rapidly growing in importance.

The active medium in a dye laser is a fluorescent organic dye dissolved in a liquid solvent. A wide range of over 200 dyes are useful for this purpose. Dyes

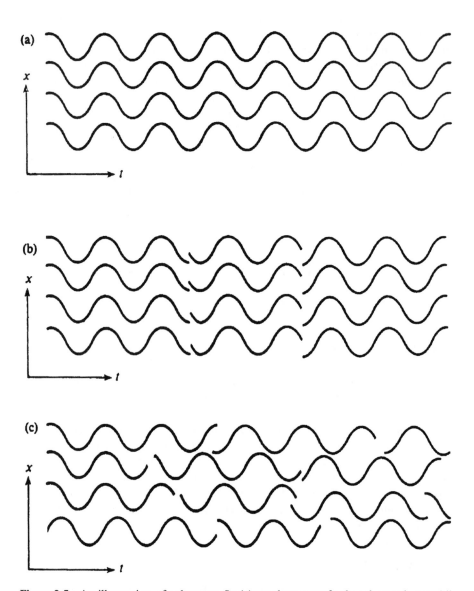

**Figure 2.5**   An illustration of coherence. In (*a*) we show a perfectly coherent beam. All the constituent waves are in phase at all times. In (*b*) we have a beam with partial spatial coherence, but which exhibits only partial temporal coherency. This is because the waves simultaneously change their phases by an identical amount every few oscillations. In (*c*) we show an almost completely incoherent beam where the phases of each wave change randomly at random times. Note, however, that even in this case some small degree of temporal coherence remains, since over very short time intervals the phases are to some extent predictable.

are large fluorescence molecules containing conjugated double bonds with multiple rings and have complex spectra. The most widely used example is rhodamine 6G, o-(6-ethylamino-3-ethylimino-2,7-dimethyl-3H-xanthen-9-yl) benzoic acid ethylester, an efficient dye with output near 580 nm.[14]

Photodegradation of the dye due to the pump light remains an inconvenient feature of these lasers. Rhodamine 6G has a lifetime of about 2000 watt hours, and many other long-wavelength dyes have lifetimes of several hundred watt hours. Solvents affect both optical properties and degradation of laser dyes. Few dyes are soluble in water, so usually they are dissolved in organic solvents such as methanol and dimethylsulfoxide. Alcohols are sometimes mixed with water to form a dye solution. The wavelength range and pump source for the major laser dyes are summarized in Table 2.1.[14]

### 2.3.1.1 Optical Properties of Organic Dyes

A simple understanding of the energy levels of a dye molecule can be obtained using the Jablonski diagram shown in Figure 2.6. Dyes can both absorb and emit light over a range of wavelengths because interactions among electronic, vibrational, and rotational energy levels create a continuum of levels in some energy ranges. The separation between vibrational levels is typically 1400–1700 cm$^{-1}$, whereas the separation of rotational levels is typically 100 times less. Optical transitions between these continua give rise to characteristic broad absorption and emission bands. The selection rules of energy transitions require that $\Delta S = 0$. Hence singlet-singlet transitions are allowed, whereas singlet-triplet transitions are forbidden.

The laser process begins with the absorption of light from an excitation source that raises the dye molecule from the ground state ($S_0$) to higher excited singlet states ($S_1, S_2, S_3, \ldots$). The laser emission occurs from the excited singlet state ($S_1$) to the ground state. A fraction of dye molecules will undergo intersystem crossing to the triplet state ($T_1$). Molecules that reach the triplet state are trapped there and are removed from the laser process. These processes contribute to a depopulation of the upper $S_1$ energy level, and a reduction in the output intensity, all of which contribute to a decrease of laser efficiency. It is most important for the laser process to avoid the intersystem crossing of molecules to the triplet state. In some cases, triplet quenchers such as dimethylsulfoxide (DMSO) are specially added to the dye solution to increase output power by repopulation of singlet states. Broadband laser emission from a dye originates from the interaction between the vibrational and electronic states of the dye molecules that splits the electronic energy levels into broad energy bands. A wavelength-selective cavity optic such as a prism or a diffraction grating can be used to tune the laser output to a desired frequency.

### 2.3.1.2 Laser Systems

Although all dye lasers have the same type of active medium, there are many different kinds of dye lasers with widely different characteristics.[15] The differ-

**Table 2.1** Major Laser Dyes

| Dye name[a] | Wavelength range (nm) | Pump source[b] |
|---|---|---|
| Polyphenyl 2 | ~383 | Short–UV Ar |
| Stilbene 1 | ~415 | UV Ar |
| Stilbene 3 | 408–453 | Nitrogen |
| (Stilbene 420)[c] | 410–454 | XeCl |
| | 412–444 | Nd:YAG, 355 nm |
| | 414–465 | UV Ar |
| Coumarin 102 | ~477 | Kr, 407–415 nm |
| (Coumarin 480) | 454–510 | Ar, ultraviolet |
| | 457–520 | Flashlamp |
| | 457–517 | XeCl |
| | 459–508 | Nd:YAG, 335 nm |
| | 453–495 | Nitrogen |
| Coumarin 30 | ~518 | Kr, 407–415 nm |
| Rhodamine 110 | 529–585 | Ar, 455–514 nm |
| (Rhodamine 560) | 529–570 | Cu, 511 nm |
| | 530–580 | Flashlamp |
| | 541–583 | Nd:YAG, 532 nm |
| | 542–578 | XeCl |
| Rhodamine 6G | 546–592 | Nd:YAG, 532 nm |
| (Rhodamine 590) | 563–625 | Flashlamp |
| | 563–607 | Cu, 511 nm |
| | 566–610 | XeCl |
| | 568–605 | Nitrogen |
| | 573–640 | Ar, 455–514 nm |
| Dicyanomethylene | 598–677 | Cu, 511 nm |
| | 600–677 | Flashlamp |
| | 600–695 | Nitrogen |
| | 607–676 | Nd:YAG, 532 nm |
| | 610–709 | Ar, 455–514 nm |
| Styryl 9 | 775–865 | Nd:YAG, 532 nm |
| | 784–900 | Ar |
| | 810–860 | Flashlamp |
| Infrared dye 140 | 866–882 | Nd:YAG, 532 nm |
| | 875–1015 | Kr, 753–799 |
| | 876–912 | XeCl |
| | 900–995 | XeCl |
| | 906–1018 | Nitrogen |

[a]The IUPAC names for the dye compounds can be found in reference 14
[b]Ar, Argon ion laser; XeCl, xenon chloride excimer laser; Kr, krypton ion laser; Cu, copper vapor laser
[c]A different name for the same compound

ences reflect both the choice of pumping source and the intended applications of the lasers. Generally, dye lasers are grouped into three major classes: laser-pumped, nitrogen, excimer, or Nd:YAG dye laser; flashlamp-pumped dye laser; and argon or krypton ion–pumped CW dye lasers.

Laser-pumped dye lasers are the simplest, most reliable, and the most easily used. Wavelength-selective optics are used in most dye lasers. There are two

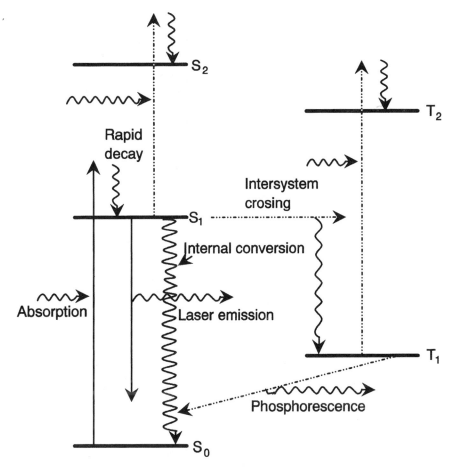

**Figure 2.6** Jablonski diagram for a laser dye.

kinds of dye laser oscillator cavity designs: the Hansch cavity and the grazing incidence cavity.

In the common Hansch configuration, shown in Figure 2.7, a wavelength-dispersive element, either a prism or a diffraction grating, is inserted inside the laser cavity. In most cases, optics are used to focus the pump light into the dye. Laser pumping can be done at different angles to the dye laser axis. The pump beam can be directed at a small angle to the axis, passing around the cavity mirrors or, in some cases, deflected around them by wavelength-dispersive tuning optics. Such longitudinal pumping can offer high efficiency, which is important for low-gain continuous-wave dye lasers. An etalon can be placed between the beam expander and the grating if a very narrow bandwidth is desired. The bandwidth of a typical Hansch dye laser is between 0.005 and 0.01 nm with just the grating and beam expander but is reduced to between 0.0005 and

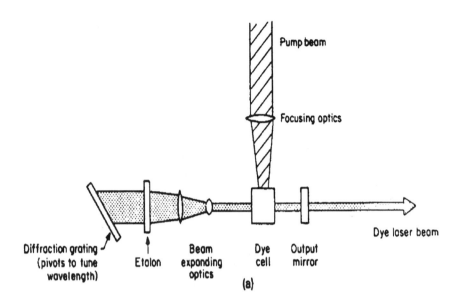

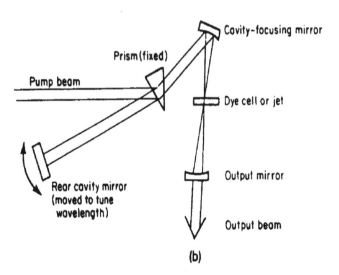

**Figure 2.7**  Typical cavity designs for dye lasers with dispersive wavelength-tuning elements. (*a*) Grating-tuned dye with pulsed pumped laser, and (*b*) prism tuning of a continuous wave dye laser.

0.002 nm with an etalon. Slightly narrower outputs can be achieved with a grazing incidence design. Typical pulse energies are 0.4, 5, and 50 mJ for nitrogen, excimer, and Nd:YAG pumped dye lasers, respectively. The pulse length and repetition rates of laser-pumped dye lasers are similar to those of the pump lasers. Nd:YAG and nitrogen lasers generate approximately 6–10 ns pulses at up to a few hundred hertz with up to several hundred millijoules pulse energy. Excimer lasers give 10–50 ns pulses at rates of up to 500 hertz with tens of millijoules energies. A flashlamp-pumped dye laser has a tube filled with a flowing dye solution that is optically coupled to a flashlamp with an elliptical cavity.[16] Flashlamp pumping gives pulses around 0.25–5 μs, with energies of 0.1–100 J, but pulses can be extended to about 500 μs. The shortest laser pulse ever recorded is 6 fs, created on a dye laser system.[17] Simon et al.[18] have demonstrated the generation of 30-fs pulses tunable over the visible spectrum. Repetition rates range from single shot to 30 hertz for coaxial flashlamp pumping, and to 100 hertz for linear flashlamps.

An example of a grazing incidence Littrow grating configuration is the typical, commercially produced dye laser model LN107 by Laser Photonics (Orlando, Florida).[19] The optical configuration of this laser is shown in Figure 2.8. The grazing incidence design provides tunability through the rotation of a mirror, and a relatively narrow linewidth can be obtained by allowing the fluorescence from the dye cell to illuminate the surface of a diffraction grating. The grazing incidence design is relatively easy to align and low cost, while the efficiencies of optical conversion are about the same for both Hansch design and the grazing incidence design, usually from 5 to 35%.

The pulse repetition rate of the emission from a dye laser is limited by the

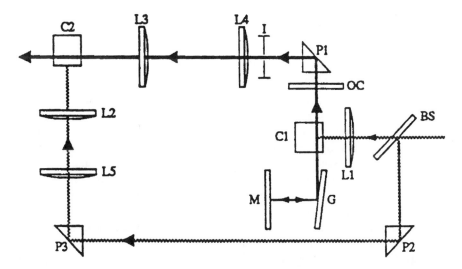

**Figure 2.8** Optical configuration of the dye laser Model LN107. C, Cell; L, lens; P, prism; OC, output coupler; BS, beam splitter; M, mirror; G, grating. From reference 19.

degradation of the dye solution and design of the dye cell. It is important to allow the dye solution to flow at an adequate rate through the dye cell, otherwise the dye solution is heated up by collisional processes that result from the very high peak energies of the pump laser passing through the dye solution. The rate of change of the volume of dye in the cell is defined by the clearing ratio, $CR$:

$$CR = r/(VP) \tag{2.5}$$

where $r$ is the flow rate of the dye solution through the dye cell, $V$ is the volume of the dye solution that is irradiated by the pump laser, and $P$ is the pump laser repetition rate. A clearing ratio of greater than two or three is required to ensure a fresh volume of dye solution to each pump laser pulse.

In practice, dye lasers that are used to produce ultrashort pulses are operated in a broadband mode without tuning optics. Synchronous mode locking can generate pulses as short as a few hundred femtoseconds, and these pulses can be further shortened with a saturable mode locker.

### 2.3.2 Gas Lasers

In gas lasers, the active medium is one gas or a mixture of different gases, which covers a wide variety of devices that dominate any list of commercially available types. Laser emission from these lasers has been seen for more lines than have been observed in other media. The basic elements of a gas laser are: a tube that can be filled with the desired gas mixture, a pair of resonator mirrors for the proper wavelength, and a suitable excitation source. Gas lasers can operate in both a pulsed and a continuous-wave mode, and can be divided into three main categories: atomic, ionic, and molecular, dependent on the laser's transition. The He-Ne, ion, molecular-gas, and excimer lasers define four of the most important classes of gas lasers in commercial use. The gas lasers cover the spectrum from the ultraviolet to the far infrared (Table 2.2).

#### *2.3.2.1 He-Ne Laser*

The helium-neon laser (Figure 2.9) has a red emission at 632.8 nm with power outputs on the order of milliwatts. The He-Ne laser is discharge-pumped, although radio frequency (RF) excitation is also possible. The active medium consists of a glass tube filled with a mixture of helium and neon gases. When an electrical current passes through helium, electrons from the discharge collide with the more numerous helium atoms and excite these atoms to a higher metastable state. Then, excited helium atoms collide with unexcited neon atoms through resonant collisions. The excited helium atoms transfer their energy to the neon atoms, raising them to a metastable state nearly identical in energy to the excited helium atoms. From the metastable state of neon atoms, electrons can return to the ground state via several paths. However, the overwhelming majority of He-Ne lasers are designed to favor 632.8 nm emission.

Most He-Ne lasers oscillate in a single transverse mode, producing a $TEM_{00}$

**Table 2.2**  Primary Wavelengths of Gas Lasers

| Type | Wavelength (nm) | Type | Wavelength (nm) |
|------|-----------------|------|-----------------|
| **Atomic** | | **Molecular** | (many lines) |
| He-Ne | 544 | CO | 5000–65,000 |
| | 594 | $CO_2$ | 9000–11,000 |
| | 612 | $N_2O$ | 10,300–11,000 |
| | 633 | | |
| | 730 | **Excimer** | |
| | 1150 | $F_2$ | 157 |
| | 1523 | ArF | 193 |
| | 3390 | KrCl | 222 |
| Xe-Ne | 2026 | KrF | 249 |
| | 3506 | XeCl | 308 |
| | | XeF | 350 |
| **Ion** | | | |
| Ar-ion | 275–477 (many lines) | **Metal-vapor** | |
| | 488 | Gold-vapor | 628 |
| | 514.5 | Copper-vapor | 510 |
| Kr-ion | 413.1 | | 578 |
| | 468 | | |
| | 520.8 | **Other** | |
| | 530.9 | $N_2$ | 337 |
| | 647.1 | HeCd | 325 |
| Ne-ion | 330–380 (many lines) | | 442 |
| Xe-ion | 526 | | 534 |
| | 535 | | 538 |
| | 540 | Iodine | 1315 |

beam with Gaussian-intensity distribution. A typical gain curve of a He-Ne laser is about 1.4 GHz, equivalent to 0.0019 nm in the red. Short, low-power models may have a small beam diameter (0.34 mm) and larger divergence (2.4 mrad). For longer, high-power models, beam diameters can reach a couple of milli-meters, with divergence about 0.5 mrad. The He-Ne laser is perhaps the most ubiquitous laser. It is a simple, low-power, low-cost, and highly reliable laser that is convenient to use. The He-Ne laser is used in low-power applications such as product-code scanning, alignment, and interferometry because of its low gain and efficiency (0.01–0.1%). Analytical applications are limited to optical alignment of other systems, or to specific experiments that are compatible with the single wavelength and low power.

### 2.3.2.2 Noble Gas Ion Laser

The noble gas ion lasers, principally argon and krypton, produce high-power outputs at many wavelengths in the visible region. Like the He-Ne laser, these gas ion lasers are typically pumped by a high-voltage dc electric discharge with current that is high enough to ionize the gas, typically several kilowatts. The excitation in an argon laser involves two steps: collisions with electrons take

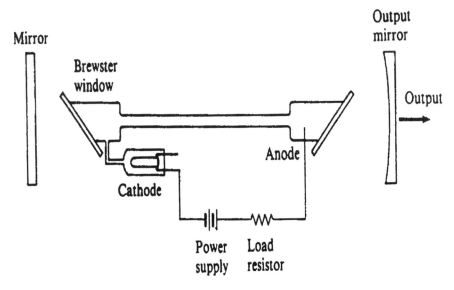

**Figure 2.9**  Schematic construction of a low-power gas laser such as the helium-neon laser.

argon atoms $(3s^23p^6)$ to the argon ion state $(3s^23p^5)$, and then other electron collisions excite outer electrons of the argon ion to 4p states. The most important visible emission lines of argon appear in the blue (488 nm) and green (514.5 nm), emitted as the result of transitions from the singly ionized states with electron configuration $3s^23p^44p^1$ down to the $3s^23p^44s^1$ state, while krypton has strong red (647.1 nm) as well as weaker yellow, green, and violet lines. Krypton generates only 10–30% as much as power as argon in the same tube. Figure 2.10 shows the energy level diagram for the argon ion laser.

Typical spectral bandwidth of a single line in an ion laser is about 5 GHz, or roughly 0.004 nm. The values of beam diameter and divergence of ion lasers are 0.6 to 2 mm and 0.4 to 1.2 mrad, respectively. Argon lasers are fairly expensive and comparatively fragile. The other major disadvantage is the plasma tube which generally has a lifetime limited to between 1000 and 10,000 hours. In analytical applications, argon and krypton lasers are used extensively for pumping CW dye lasers and mode-locked dye lasers. The visible argon ion laser also fits the requirements of many measurement applications including laser Doppler velocimetry, laser Doppler anemometry, particle sizing, and so forth.

### 2.3.2.3  Carbon Dioxide Laser

The carbon dioxide laser is one of the most powerful lasers. The energy levels involved in carbon dioxide laser action are rotational and vibrational levels, and emission therefore occurs at much longer wavelengths, well into the infrared region. The energies of three vibrational modes, symmetrical stretching ($\upsilon$ =

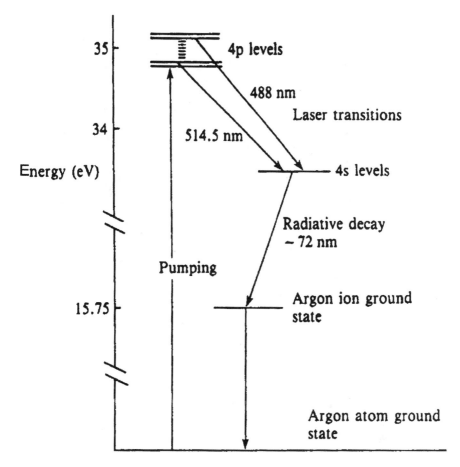

**Figure 2.10** Simplified energy-level diagram for the argon laser.

1337 cm$^{-1}$), asymmetrical stretching ($\upsilon = 2349$ cm$^{-1}$), and bending ($\upsilon = 667$ cm$^{-1}$), correspond to around 10 μm. The carbon dioxide gas is usually accompanied by nitrogen and helium. Most carbon dioxide lasers are discharge-pumped, but RF excitation is used also. The discharge is pulsed at a rate of about 20 per second and produces output of hundreds of kilowatts. Oscillation occurs between two vibrational levels in carbon dioxide, while the nitrogen and helium greatly improve the efficiency of laser action. The output power of the laser can be increased by use of the Q-switch technique (Section 2.4).

From the point of view of their construction, carbon dioxide lasers can be separated into six categories: (1) longitudinal flow lasers, (2) sealed-off lasers, (3) waveguide lasers, (4) transverseflow lasers, (5) transversely excited atmospheric pressure (TEA) lasers, and (6) gas-dynamic lasers. The most efficient is the TEA laser (Figure 2.11). This laser's prime attractions are the generation of

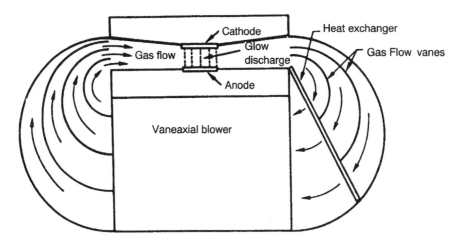

**Figure 2.11**   Schematic diagram for a transverse-flow carbon dioxide laser.

short, intense pulses, up to megawatts of peak power, and the extraction of high power per unit volume of laser gas.

Maximum output depends on the type of carbon dioxide laser. Waveguide lasers can produce continuous-wave output up to about 50 W, while sealed carbon dioxide lasers can reach about twice that level. Longitudinal flow lasers can reach the kilowatt range, while transverse-flow lasers can produce several kilowatts. Average powers of TEA lasers range up to a few thousands watts.

The 10.6-$\mu$m emission of carbon dioxide is ideal for drilling, cutting, and welding a variety of materials for industry, because of good gain and very high efficiency (up to 30%). Other applications include isotope separation, laser pyrolysis, and laser enhancement of chemical reactions.

### 2.3.2.4 Nitrogen Laser

A nitrogen laser is normally a transverse discharge device through which gaseous nitrogen flows. This laser is relatively inexpensive, very reliable, and has low operating costs. As one of the first ultraviolet lasers on the market, it played a key role in the spread of pulsed lasers. The 337-nm wavelength of the nitrogen laser can be used to pump most laser dyes. The output of this laser is typically a few hundred kilowatts for a few nanoseconds and is energetic enough to pump dyes easily with use of inexpensive optics and dye cells for emission over the 360–950 nm range. A megawatt nitrogen laser is commercially available.[20]

Figure 2.12 shows the energy-level diagram for neutral nitrogen molecules involved in the 337.1-nm laser transition. A fast, high-voltage discharge excites the nitrogen molecules, populating the upper laser level—an excited electronic state with 40-ns lifetime. The excited nitrogen molecules emit at 337.1 nm when they drop to the lower laser level. The transition is a vibronic state in which both vibrational and electronic energy levels change, making it a broadband by

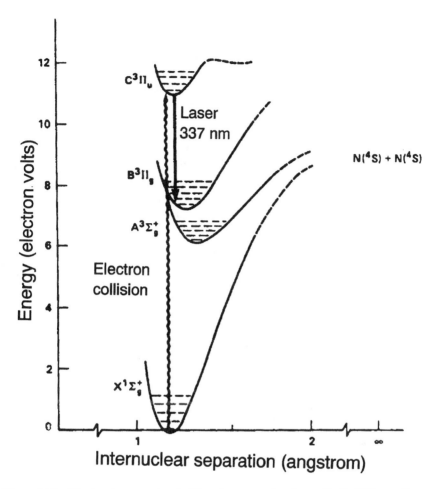

**Figure 2.12**  Electronic energy level of the nitrogen molecule used in the 337-nm nitrogen second-positive-band laser.

laser standards. The rapid population and quick decay of the upper laser level combine to give nitrogen lasers very high gains during their short pulses, typically 50 dB/m of cavity length or more.

The output of commercial nitrogen lasers ranges from tens of microjoules to less than 10 mJ per pulse. Duration and repetition rates of nitrogen laser pulses are dependent on gas pressure, discharge circuitry, and internal kinetics. Typical pulse duration ranges from about 300 ps at atmospheric pressure to about 10 ns at 20 torr, and repetition rates are 1–100 Hz.

### 2.3.2.5  Excimer Laser

Excimers are molecules that are formed and exist only in the electronically excited states. Molecules returning to the ground state emit short wavelength radiation. Excimer lasers are powerful and efficient gas lasers that utilize electronic transitions in rare gas halide (RGH) molecules to produce intense pulses of ultraviolet (UV) light. Formed by gas-phase chemical reactions in an electrical discharge, these diatomic RGH species are bound in the electronically excited upper laser level, but dissociate upon reaching the ground state, lower laser level. This leads to nearly complete regeneration of the original fuel gases in the laser after each discharge, allowing many output pulses to be produced from a single fill of the laser gas reservoir. The most important emission lines are: argon fluoride (ArF), 193 nm; krypton fluoride (KrF), 248 nm; xenon fluoride (XeF), 354 nm; argon chloride (ArCl), 175 nm; krypton chloride (KrCl), 222 nm; xenon chloride (XeCl), 308 nm; and xenon bromide (XeBr), 282 nm.

Excimer laser beams are not round with a Gaussian profile. Most excimer beams are rectangular and have a top hat spatial profile, which is a flat top with steeply sloped sides in one dimension, usually the longer dimension. The other dimension may be near-Gaussian, or top hat, or something in between. These rectangular beams are suitable for large area materials processing applications that require a uniform energy density at the workpiece. Figure 2.13 shows a laser beam intensity profile.[21]

The mechanism for selectively populating the excimer levels can, in a simplified sense, be viewed as a sequence of collisional energy exchanges. The active medium of the excimer laser is an excited diatomic complex rich in inert gas halides such as ArF, KrF, XeCl, and so on. The diatomic complex is formed by a chemical reaction between an inert gas and halide ions, which is produced by the electrical discharge. The electrical discharge excites the diatomic complex to a higher excited state. This state has a potential energy minimum, allowing the population inversion to occur. The chemical reactions are described as follows:

1. Ionization process

$$Ar + e^- \rightarrow Ar^+ + 2e^-$$

$$F_2 + e^- \rightarrow F^- + F$$

2. Diatomic complex formation process

$$F^- + Ar^+ + He \rightarrow (ArF)^* + He$$

3. Lasing process

$$(ArF)^* \rightarrow Ar + F + h\nu$$

4. Molecular reformation process

$$F + F \rightarrow F_2$$

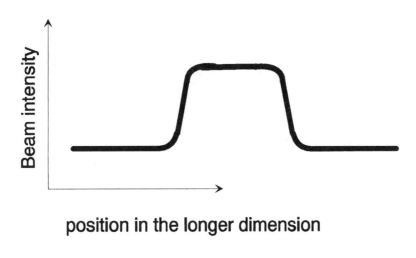

position in the longer dimension

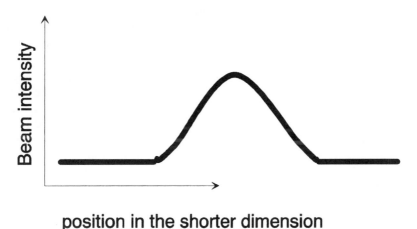

position in the shorter dimension

**Figure 2.13** Excimer laser beam temporal intensity profile; horizontal profile (*top*), vertical profile (*bottom*). From reference 21.

Among the above chemical reactions, helium acts only as a buffer gas, without direct chemical reaction involved. The buffer provides an environment from which the complex can be formed. If there are some impurities in the laser cavity, $Ar^+$ or $F^-$ may react with these impurities instead of forming $(ArF)^*$. Therefore, high purity laser gases are critical.

The gas-handling system is an important consideration in the operation of excimer lasers. Operation cost is high since expensive high-purity rare gases are required and must be changed periodically, although the rare gases can be cleaned and recycled after the laser gas mixture degrades during operation.

The wide output energy range and short wavelengths of excimer lasers make them attractive for applications that range from high-power industrial to medical, and from spectroscopic to micromachining. The average power starts at a few milliwatts in waveguide excimers and has recently reached 1 kW in an experimental X-ray preionized device.[22]

### 2.3.3 Semiconductor Laser

Diode lasers based upon the various semiconductor materials have proved to be unique tools in molecular spectroscopy as well as atomic spectroscopy.[23–25] Both the II–VI materials group, PbSeTe, and III–V materials group, GaAs, are used, as the fundamental physics of these two groups of materials are the same.

The fundamental physics of these lasers is sketched very briefly here. More specialized reviews should be consulted for further details.[26–28] The main difference from other lasers is that the energy levels in semiconductors must be treated as continuous distributions of levels rather than as discrete levels, as in other laser systems. The energy-level scheme for an idealized semiconductor is shown in Figure 2.14.

The energy-level spectrum consists of very broad bands. There is a small energy gap between two energy bands known as the valence band, v, and the conduction band, c. Each band actually consists of a large number of very closely spaced energy states. Semiconductor diode lasers are excited by current passed through a rectifying junction.[28] A semiconductor laser contains both p-type, e-acceptor, and n-type, e-donor, materials. Such a device acts as a diode, so a semiconductor laser is also called a diode laser. By applying an electrical potential in forward bias mode across a diode junction, electrons crossing the semiconductor boundary drop down from a conduction band to a valence band,

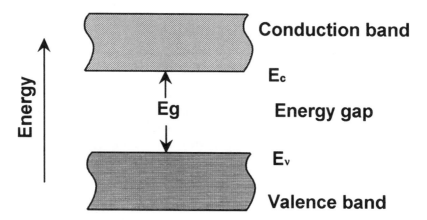

**Figure 2.14**  The energy band structure of a semiconductor.

emitting radiation by an electron-hole recombination process. The emitted wavelength is given by the following equation:

$$hc/\lambda = E_c - E_v = E_g. \tag{2.6}$$

The wavelength of the diode laser can be tuned either by diode current over a short spectral region of about 3 nm or by temperature over a wider range of about 20 nm for a temperature change of $-20$ to $+60$ °C. The tunability by diode current is due to the change of the index of refraction in the active region when the current density is changing. The temperature tunability is caused by a change in the effective band gap at the n-p junction. Wavelengths of the diode laser cannot be tuned continuously by current or temperature because of mode hopping.[25] Some of the important semiconductor laser light sources available are given in Table 2.3, and the typical performance of a diode laser is shown in Table 2.4.

The wavelength of the diode laser can sometimes be extended to shorter wavelengths by second harmonic generation. The diode laser has other advantages over conventional lasers. It is smaller and less expensive, easy to operate, and it has a long lifetime, about $10^5$ hours. Furthermore, it has a high conversion efficiency from electricity to light and can be electrically driven and controlled.

**Table 2.3**  Important Semiconductor Laser Materials and Emitting Wavelengths

| Compound | Wavelengths (nm) | Notes |
| --- | --- | --- |
| ZnSe | 525 | First demonstrated in 1991 |
| AlGaInP | 630–680 | Developmental; lifetimes have been short at shorter wavelengths |
| $Ga_{0.5}In_{0.5}P$ | 670 | Active layer between AlGaInP layers; long room-temperature lifetime |
| $Ga_{1-x}Al_xAs$ | 620–895 | $x = 0$ to 0.45; lifetimes very short for wavelengths below 720 nm |
| GaAs | 904 | |
| $In_{0.2}Ga_{0.8}As$ | 980 | Strained layer on GaAs Substrate |
| $In_{1-x}Ga_xAs_yP1-y$ | 1110–1650 | InP substrate |
| $In_{0.73}Ga_{0.27}As_{0.58}P_{0.42}$ | 1310 | Major fiber-communication wavelength |
| $In_{0.58}Ga_{0.42}As_{0.9}P_{0.1}$ | 1550 | Major fiber-communication wavelength |
| InGaAsSb | 1700–4400 | Possible range; developmental; on GaSb substrate |
| PbEuSeTe | 3300–5800 | Cryogenic |
| PbSSe | 4200–8000 | Cryogenic |
| PbSnTe | 6300–29,000 | Cryogenic |
| PbSnSe | 8000–29,000 | Cryogenic |

**Table 2.4** Typical Performance of a Commercial Diode Laser

| Parameter | Performance | Notes |
|---|---|---|
| Output power | 30–40 mW | Continuous |
| | 10 W | Pulsed |
| Wavelength | 670–850 nm | Fundamental |
| | 420 nm | second harmonic generation, 0.4 mW |
| Linewidth | 1 nm | Conventional |
| | 10 MHz | Stabilized |
| Tunable range | 20 nm | Temperature controlled |
| Pulse width | 0.1–100 ns | No mode locking |
| Stability | 0.006% | Controlled |
| Efficiency | 10–20% | Continuous wave |

However, the small size, usually about half a millimeter, does result in very poor collimation, very wide beam divergence (typically $\theta = 40°$), and thus creates the need for corrective optics in many applications.

The use of a tunable semiconductor laser as a light source for multi-elemental atomic absorption analysis has been reported by Hergenröder and Niemax,[29–31] and Winefordner et al.[32] Lower detection limits and wider linear dynamic ranges were obtained. The major disadvantage of using the semiconductor laser in atomic absorption spectrometry is the limited wavelength range (660–860 nm). However, the introduction of laser diodes with fundamental wavelengths below 660 nm will give a more versatile system. Readers are referred to several recent reviews on the applications of tunable diode lasers in high-sensitivity spectroscopy.[23–25]

### 2.3.4 Solid-State Lasers

Solid-state lasers include all optically pumped lasers in which the gain medium is a solid at room temperature. The active medium in solid-state lasers is generally a transparent crystal or glass which is doped with a small amount (less than 1%) of transition metal. The two most common dopant metals are chromium, in the ruby and alexandrite lasers, and neodymium, in the Nd:YAG and Nd:glass lasers.

### 2.3.4.1 Ruby Laser

The ruby laser was invented first and is still important. Population inversion is obtained in a ruby laser by optical pumping. The ruby is a single crystal of aluminum oxide ($Al_2O_3$) within which a small proportion of the aluminum ions are replaced by $Cr^{3+}$ ions. The chromium ions in ruby are excited by broadband emission from a flashlamp filled with xenon and coiled around it or placed alongside it within an elliptical reflector. The ruby laser is defined as a three-level laser because the electronic transitions in chromium ions consists of three energy levels (Figure 2.15).

The $Cr^{3+}$ ions in the ruby crystal are excited first into the intensely absorbing $^4F_1$ and $^4F_2$ bands from which they quickly relax to the $^2E$ excited state. A population inversion is established between $^2E$, which has a much longer lifetime (3 ms), and the ground state $^4A_2$. The stimulated emission occurs at 693.4 nm. High energy is required for population inversion in ruby because the laser transition terminates in the ground state and, therefore, a large number of electrons must be pumped out of the ground state. A Q-switch is often used in the cavity to control the output by concentrating all of the energy into a single, intense pulse with a duration of about 25 ns. A single pulse can have 10 J or more of energy.

The duration of the heating and cooling of the ruby rod normally restricts the pulse repetition rate. Some form of active cooling is necessary to improve the performance of the ruby laser. Typical designs circulate deionized water through the laser head and through a heat exchanger.

A common method of operating the ruby laser is to use an oscillator-amplifier combination. The output from an oscillator is passed through a second flash-lamp-pumped ruby rod housed in another elliptical cavity. The output power is amplified in a single pass through the second rod and can be greater than 100 J. This configuration keeps the intracavity powers lower in both the oscillator and amplifier so the combination is less sensitive to optical damage.

The typical bandwidth of the emission is about 0.5 nm, but this can be reduced by a factor of up to 100 by use of intracavity etalons. The beam diameter

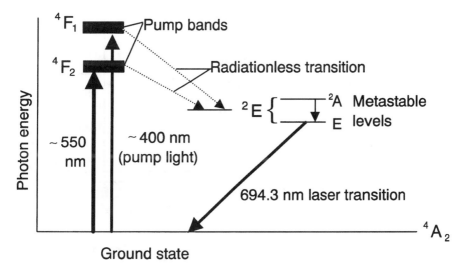

**Figure 2.15** Energy levels of $Cr^{3+}$ in ruby laser show pump and laser transitions.

of a low-power ruby laser can be as little as a 1 mm, with 0.25 milliradian divergence.

### 2.3.4.2 Neodymium Lasers

Neodymium lasers are of two main types: neodymium-doped glass (Nd:glass) and neodymium-doped yttrium aluminum garnet (Nd:YAG) lasers. In 1961, Snitzer[33] demonstrated an Nd:glass laser using barium crown glass. After this, other rare earth–doped glasses were made into lasers; dopants included Yb, Ho, Er, Tm, and Tb.[34-38] By 1963, rare earth–doped laser glass did not seem to be competitive with crystalline YAG because the glass has lower thermal conductivity than YAG.

Although transitions in the neodymium ions are responsible for laser action in both cases, the emission characteristics differ because of the influence of the host lattice on the neodymium energy level. In Nd:YAG, high pumping efficiencies can be obtained from the four-level laser system, as shown in Figure 2.16. Optical pumping raises the neodymium ions from the ground state to a higher energy level, from which there is decay to a metastable level, which in

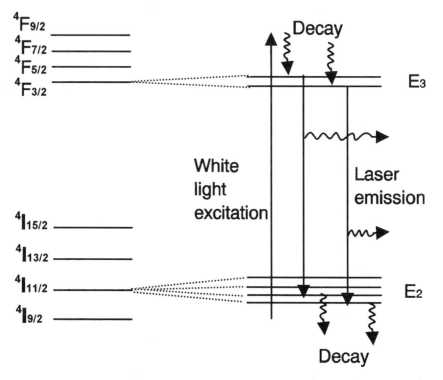

**Figure 2.16** Energy-level diagram for the neodymium ion in an Nd:YAG laser, showing the principal laser transition.

turn produces a population inversion between the $^4F_{3/2}$ and $^4I_{11/2}$ states. It is much easier to achieve a population inversion in Nd:YAG because the latter state ($^4I_{11/2}$) is sufficiently far above the ground state to be practically empty at room temperature. The initial and final states for lasing action are split into 2 and 6 crystal field levels, respectively, so that several lasing wavelengths are possible. The most powerful of these occurs at 1.064 μm, and this is usually the one used. However, the majority of Nd:YAG research applications in analytical spectroscopy do not use the wavelength of 1.064 μm directly. Particularly useful wavelengths are 532, 355, and 266 nm obtained by harmonic generation (see Section 2.4).

The output energy at 1.064 μm from YAG lasers ranges from a tenth of watt to hundreds of watts, and the lasers can be operated in a single ($TEM_{00}$) mode, in multiple transverse modes, or in modes determined by an unstable resonator. Pulsed output can be obtained from continuously pumped Nd:YAG lasers by Q-switching or mode locking. The divergence of Nd:YAG laser beams can be from a fraction of a milliradian to around 10 mrad, and beam diameter is 1–10 mm.

The Nd:glass laser produces laser radiation at 1.053 μm. Glass has several advantages as a host material because it is isotropic, cheaply and easily fabricated, and can be doped with a large concentration of neodymium to produce higher power than Nd:YAG. On the other hand, it has a lower thermal conductivity which prohibits CW operation and, indeed, limits the pulse repetition rate in pulsed operation.

### 2.3.4.3 Tunable Solid-State Lasers

Tunable solid-state lasers date back to the first demonstration of the Ni:MgF$_2$ laser by Johnson et al.[39] in 1963. The discovery and subsequent development of color center lasers[40] and the alexandrite (Cr:BeAl$_2$O$_4$) laser[41] provided encouraging signs that broadly tunable devices based on solids, rather than liquids, might be available. These crystalline media can be wavelength tunable because of the vibronic nature of the laser transition. A vibronic transition involves simultaneous emission of an optical photon and excitation of a vibrational mode of the crystal. Figure 2.17 shows an energy-level diagram for the alexandrite laser. Ions are optically excited by the pump radiation to a broad continuum of high-lying states ($^4T_1$ or $^4T_2$) which rapidly decay to the closely spaced storage level ($^2E$). Electrons in the storage level are thermally excited into the $^4T_2$ level, and the lasing transition occurs from this level to one of the many vibronic levels. The high density of vibrational ground states in vibronic laser systems allows partitioning of the total emitted energy between photons, and results in optical gain over a broad continuum of wavelengths over the range of 700–815 nm. The alexandrite laser has visible absorption bands in the blue and red spectral regions, and therefore can be pumped by xenon flashlamps or red diode lasers.

The most commercially successful vibronic laser is the Ti:sapphire (Ti:Al$_2$O$_3$)

laser which has emerged as a solid-state alternative to the dye laser.[42] Its popularity can be attributed to a very broad spectral output range (670–1070 nm) and excellent mechanical and thermal properties. The Ti:sapphire laser also has high gain and can be run either CW or pulsed. The gain bandwidth of the Ti:sapphire laser, as shown in Figure 2.18, is extremely large. The peak gain

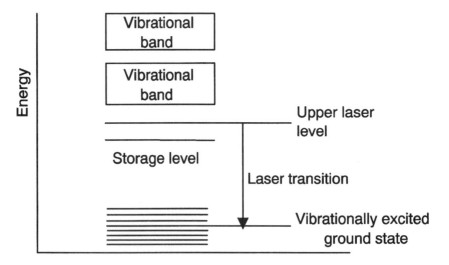

**Figure 2.17**    Energy-level diagram for the alexandrite laser.

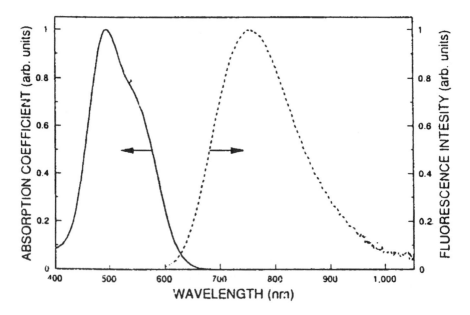

**Figure 2.18**    Absorption and emission spectra of room-temperature Ti:sapphire.

cross section of approximately $3.5 \times 10^{-19}$ cm$^2$ is comparable to that of the Nd:YAG.[42] The general application of the Ti:sapphire laser includes high-resolution linear and nonlinear spectroscopy of atoms, molecules, and condensed media.

The optical parametric oscillator (OPO) laser appears to have a promising future in analytical chemistry. The first successful research and theories of OPO were reported in 1965.[43] However, commercialization of OPO technology for the visible and UV regions occurred only recently when suitable nonlinear optical materials become available. The ideal crystals for an OPO have to possess the following properties:[44-47] (1) high damage threshold to sustain the intense pump fluence required for the nonlinear interaction; (2) phase matching conditions for the pump, "signal," and "idler" wavelengths over a large tuning range; (3) ability to be made in useful size; (4) chemically stable, with no significant degradation over time; and (5) low absorption over the entire tuning range. Some properties of several nonlinear optical crystal materials have been summarized by Simon and Tittel.[48] Several nonlinear crystals such as $\beta$-BaB$_2$O$_4$ (BBO), LiB$_3$O$_5$ (LBO), and KTiOPO$_4$ (KTP) show promise as OPO media.

A parametric interaction[47,49,50] between the pump light and the molecules of the crystal occurs when laser light of sufficient intensity is focused onto a nonlinear birefringent crystal. The pump photon at frequency $\omega_p$ is split into two photons: the signal, at frequency $\omega_s$, and the idler, at frequency $\omega_i$. In order to maintain energy conservation, the relationship $\omega_p = \omega_s + \omega_i$ holds. Momentum conservation demands that $K_p = K_i + K_s$ for the corresponding wave vector for pump, signal, and idler waves. The above frequency relationship can also be expressed in terms of wavelength, $1/\lambda_p = 1/\lambda_s + 1/\lambda_i$. Coherent superimposition will result from phase matching if the refractive index of the crystal has the proper value for $\omega_p$, $\omega_s$, and $\omega_i$ in the directions of $K_p = K_i + K_s$. This gives high-intensity coherent waves with frequencies $\omega_s$ and $\omega_i$. The tunable output at $\omega_s$ and $\omega_i$ can be obtained by rotation of the nonlinear crystal.

Ebrahimzadeh et al.[49] studied a $\beta$-BaB$_2$O$_4$ (BBO) optical parametric oscillator pumped at 308 nm by a narrowband injection-seeded ultraviolet XeCl excimer laser. The pulsed, singly resonant OPO provided tunable coherent radiation over the wavelength range from 354 nm in the near UV, through the visible, to 2.370 $\mu$m in the near infrared. The measured linewidths of the signal waves (approximately 1.5–10.5 nm) in wavelength units (nanometers) were narrower than the corresponding idler linewidths (approximately 2.5–12.5 nm) across the oscillator tuning range. Optical-to-optical energy conservation efficiencies of about 10% were obtained by use of a 12-mm-long crystal in the wavelength range from about 454 to about 958 nm. With an input pump pulse energy of 17.5 mJ, output energies approaching 2 mJ per pulse were obtained from the OPO. The tuning range of this OPO could be extended further into the near ultraviolet (down to about 270 nm) by use of other excimer lasers, such as KrF at 249 nm, as the pump source.

Figure 2.19 is a diagram of a 355-nm, third harmonic Nd:YAG pumped OPO laser system, based on a commercial laser design.[51,52] The third harmonic output

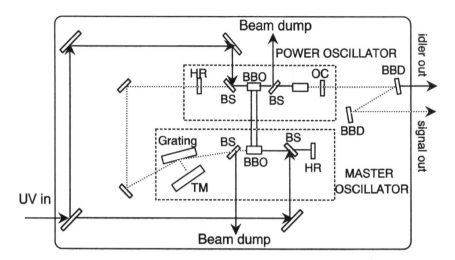

**Figure 2.19** MOPO optical schematic diagram. HR, High reflector; BS, beam splitter; BBO, crystal; OC, output coupler; BBD, broadband dichroic mirror; TM, tuning mirror. From reference 51.

energy generated by the Nd:YAG laser is approximately 600 mJ. The repetition rate of the laser is 10 Hz, and the pulse duration is typically 6 ns. The output linewidth is about 0.2 cm$^{-1}$ across the tuning range (410–2000 nm). Output energies in excess of 100 mJ per pulse in the visible (signal) region, and over 50 mJ per pulse in the near IR (idler), range have been achieved with this OPO system. The maximum frequency-doubling efficiencies of the output of this laser system are 18.5% when using BBO and 28% with KDP. Accordingly, output energies in the UV are from 10 to 20 mJ per pulse over the central portion of the tuning range (about 240–345 nm) and 5–10 mJ per pulse at either end (approximately 215–240 nm and 365–450 nm). The extremely broad tuning range, ease of operation, high conversion efficiency, and the potential high output power should make the OPO a very competitive laser in analytical chemistry.

## 2.4  Laser Instrumentation

### 2.4.1  Frequency Conversion

An important region of the spectrum for atomic spectroscopy, between 195 and 320 nm (the UV region), can not be directly obtained with either dye lasers or any other type of laser. However, it can be obtained through optical frequency conversion techniques, that is, frequency doubling, sum or difference frequency generation, and Raman shifting.

The most common technique is second harmonic generation (SHG), commonly called frequency doubling. Consider the following equation

$$P = \varepsilon_0(\chi_1\varepsilon + \chi_2\varepsilon^2 + \chi_3\varepsilon^3 + \ldots) \tag{2.7}$$

where $\chi_1$ is the linear susceptibility of the crystal, and $\chi_2$, $\chi_3$, and so forth are the nonlinear optical coefficients. If we substitute $\varepsilon = \varepsilon_0 \sin \omega t$, we can obtain

$$P = \varepsilon_0(\chi_1\varepsilon_0 \sin \omega t + \chi_2\varepsilon_0^2 \sin^2\omega t + \chi_3\varepsilon_0^3 \sin^3 \omega t + \ldots). \tag{2.8}$$

The second term may be rewritten as $1/2\ \varepsilon_0\chi_2\varepsilon_0^2(1 - \cos 2\omega t)$. Thus it is expected that some radiation at $2\omega$ may be produced. Frequency doubling was first observed by Franken et al.[53] They observed a low intensity of 347.15 nm radiation through a quartz crystal from ruby laser light. The most widely used crystals are KDP ($KH_2PO_4$), lithium niobate ($LiNbO_4$), and β-barium borate (BBO). Although several orders of the nonlinear effect can be defined, only second-, third-, and sometimes fourth-order frequency shifts are used for the most practical applications in laser spectroscopy.[54,55] Generation of the third and fourth harmonics by use of nonlinear crystals is usually a multistep process.

The commercially available electrooptic materials are ADP, $(NH_4)H_2PO_4$; KDP, $KH_2PO_4$; $KD_xP$, $KD_2PO_4$; KDA, $KH_2AsO_4$; RDP, $RbH_2PO_4$; ADA, $(NH_4)H_2AsO_4$; BBO, β-$BaB_2O_4$; GaAs; GaP; ZnTe; ZnS; CdTe; CuCl; $Bi_4(GeO_4)$; ZnO; CdS; CdSe; $LiNbO_3$; $LiTaO_3$; $LiIO_3$; $SrTiO_3$; $KTaO_3$; $BaTiO_3$; $K_3Li_2Nb_5O_{15}$; $BaSrNb_4O_{12}$; $Sr_2KNb_5O_{15}$; $BaSrNaNb_5O_{15}$; $N_4(CH_2)_6$; and Se. The commercial available quadratic electrooptic materials are $SrTiO_3$, $KTaO_3$, $BaTiO_3$, LiCl, NaCl, LiF, NaI, and NaF. The choice of material depends upon operating wavelength and power-handling requirements.

At low incident intensities, the output intensity of these crystals increases with the square of the incident intensity, while the efficiency increases linearly with incident intensity. However, at very high incident intensities, the efficiency begins to decrease as a result of a competing absorption process within the crystal.[56] The optical efficiency of most commercial crystals ranges from 10 to 40%. In order to obtain high conversion efficiencies, the incident radiation can be focused into the doubling crystal by means of a lens. If the radiation is focused too tightly into the crystal, however, damage may result. Therefore, a compromise between UV output and crystal lifetime has to be made.

The frequency-doubled output and incident laser radiation must be phase matched, otherwise the two beams would interfere destructively.[57,58] Phase matching requires different crystal angles with respect to the input laser beam at different wavelengths. Therefore, it is important that the rotation of the crystal is synchronized with the tuning of the laser wavelength.

The second frequency conversion technique is frequency mixing, which involves the mixing of two or more primary laser beams of different frequencies within a nonlinear material. In general, if the input frequencies are $\omega_1$ and $\omega_2$, the nonlinear process generates additional waves at frequencies $\omega_1 + \omega_2$ and $\omega_1 - \omega_2$. Typically, the difference frequency is selected for use, although the physical interaction generates both.

The third frequency conversion technique involves Raman shifting.[56,59] Raman shifting relies upon the stimulated Raman scattering (SRS) effect and may be used to generate laser wavelengths in the 190–220 nm region and further into the vacuum UV. It involves the inelastic scattering of light by atoms. The scattering process results in either a gain or a loss of energy by the atoms. As a result, the frequency of the scattered light differs from the irradiation frequency, $\nu_0$. The Stokes Raman transition from level $E_1$ to level $E_2$ results in scattering of a frequency given by

$$\nu_S = \nu_0 - \Delta E/h, \tag{2.9}$$

and the corresponding anti-Stokes transition from $E_2$ to $E_1$ produces a frequency

$$\nu_{AS} = \nu_0 + \Delta E/h \tag{2.10}$$

where $\Delta E = E_2 - E_1$, and $h$ is Planck's constant. Thus the frequency of incident light shifts by $\Delta E/h$ to the negative and positive sides of the dominant Rayleigh line. Examples of commonly used Raman scattering media are hydrogen, deuterium, methane, and ammonia gases. One of the advantages is that wavelengths as short as 150 nm can be obtained if calcium fluoride windows are used in conjunction with a laser optical path that has been purged with nitrogen.

Funayama et al.[60] reported a continuously tunable coherent source from 202 to 3180 nm based on a Ti:sapphire laser. Nonlinear conversion techniques, such as harmonic generation in BBO crystals and SRS in high-pressure gases (hydrogen or methane), were used to extend the tunable range. In most of the spectral regions, the output energy exceeded 3 mJ per pulse. Further extension of the tunable range from 120 nm to 15 $\mu$m would be possible by the use of Stokes and anti–Stokes Raman shifting.

## 2.4.2 Q-Switching and Cavity Dumping

In laser atomic spectroscopy, high pulse energies of lasers are desired. Both Q-switching and cavity dumping techniques are used to provide very high laser pulse energies.

In a Q-switched laser, a shutter or some other device is placed within the cavity. During the off cycle, no appreciable stimulated emission takes place and a sizable population inversion is established in the active medium. When the shutter is opened such that the cavity becomes a resonator, the energy stored by the medium is quickly released in a pulse of intense radiation. This technique involves first reducing then swiftly increasing the Q-factor of the laser. The Q factor is defined as the ratio of the emission frequency to the linewidth,

$$Q = \nu/\Delta\nu \tag{2.11}$$

or as

$$Q = 2\pi \text{ energy stored/energy loss.} \tag{2.12}$$

Normally, the Q factor of a laser cavity is constant, but modulating the Q factor raises interesting possibilities. If the Q factor is kept artificially low, say by putting a lossy optical element into the cavity, energy will gradually accumulate in the laser medium because the Q factor is too low for laser oscillation to occur and dissipate the energy. If the loss is removed suddenly, the result is a large population inversion in a high-Q cavity, producing a high-power burst of light a few nanoseconds to several hundred nanoseconds long in which the energy is emitted. This rapid change in cavity Q is called Q-switching. A Q factor of $10^7$ is not uncommon in the lasers most widely used for spectroscopy.

In cavity dumping, the energy within the cavity is allowed to build up with fully reflective end mirrors until sufficient population inversion is established. The beam is then directed out of the cavity by a third coupling mirror or an acoustooptical modulator. All of the energy in the beam can be dumped out in a single pulse. Laser action is terminated immediately and does not recommence until the coupling mirror is switched out of the beam or the modulator is back to normal (Figure 2.20).

Both cavity dumping and Q-switching are methods of storing energy to increase the output power. In Q-switching, energy is stored in the form of excited states, while in cavity dumping the energy is stored in the form of intracavity power. Q-switching is applicable for laser materials having long-lived excited states, such as ruby and Nd:YAG crystals, whereas cavity dumping is applicable to materials with short excited lifetimes, such as argon ions and dyes.

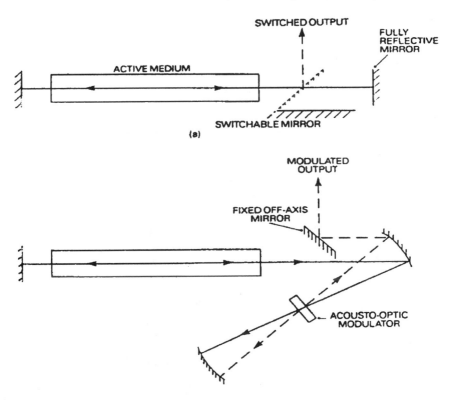

**Figure 2.20**  Schematic diagram of typical cavity dumping.

### 2.4.3 Ultrashort Pulse Techniques

Ultrashort laser pulses (in the femtosecond to picosecond range) are needed for ultrafast laser spectroscopy,[61] in which the fast time scale extends approximately from 10 fs to 100 ps. Mode locking[62,63] is one of the ultrashort pulse techniques used to produce laser pulses of ultrashort duration, typically picoseconds, and very high peak power, typically a few gigawatts. This can be accomplished by modulating the loss (or gain) of the laser cavity at a frequency equal to the intermode frequency separation $\Delta v = c/2L$, where $L$ is the length of the resonator. Typical $\Delta v$ is around $10^8$ Hz. This technique creates a phase relationship such that completely constructive interference between all modes occurs at just one point, and destructive interference everywhere else. The result is the generation of a repetitive train of pulses at the output.

There are several ways to accomplish mode locking. One method is to modulate the gain of the laser cavity by placing an electro-optical or acousto-optical switch inside the cavity, much the same as a Q-switching, but driving it at the beat frequency of adjacent modes (i.e., $c/2L$). The device becomes transparent for brief intervals separated by $\tau = 2L/c$. Hence, only electromagnetic waves that travel in phase with the modulation will be amplified, which results in phase locking of the modes. Other mode-locking techniques, such as synchronous pumping, are mostly used for tunable dye lasers. The pulse train envelope of a synchronously pumped mode-locked dye laser has been experimentally studied as a function of cavity length detuning.[64]

Although mode locking produces a train of pulses, it is possible to select out single pulses by using either a Pockels cell-polarizer combination placed outside the cavity, or the cavity dumping technique. Cavity dumping can enhance the pulse intensity many times higher than can be obtained by mode locking alone. It is commonly employed to generate a picosecond pulse train of variable repetition rates.

Recent advances in ultrashort pulse generation techniques suggest that a new ultrashort pulsed laser technology can be developed for solid-state lasers. One method for developing an ultrashort pulse mode-locked solid-state laser is the use of nonlinear optical feedback.[65,66] The resultant laser is called a soliton laser, as it is based on the use of single-mode fibers to form solitons, the self-maintaining pulses whose shapes are periodic with propagation. The device is shown in Figure 2.21. It consists of a mode-locked color center laser with a nonlinear external cavity. Output pulses from the main laser are coupled into an optical fiber. With the addition of the fiber arm, the device operates as follows:[65] As the laser action builds up from noise, the initially temporally broad pulses are considerably narrowed by pulse compression and pulse splitting when they pass through the optical fiber. The narrowed pulses are reinjected back into the main cavity and force the laser itself to produce narrower pulses. This process builds upon itself until the pulses in the fiber become solitons, that is, until the pulses have substantially the same shape following their double passage through the fiber as they had upon entry. Unlike most mode-locked lasers, the soliton laser

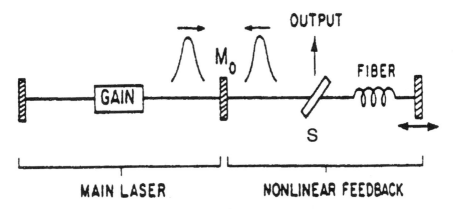

**Figure 2.21**  Schematic diagram of the ultrashort pulse generation using nonlinear optical feedback. Mo, Output coupler; S, beam splitter.

can be made to have any desired pulse width over a wide range from several picoseconds to under 100 femtoseconds, through the choice of the length of fiber used in the feedback loop.

For many mode-locked lasers, the mode-locking process and the factors that control it are not yet completely understood. It has been argued that the soliton laser may be only one example of a much wider class of systems in which the external cavity has some nonlinear response but does not necessarily support solitons. For example, Blow et al.[67,68] reported improved mode locking with a nonlinear nonsoliton external cavity. Their experiments concluded that solitons are not essential for the operation of a soliton laser,[68] and that mode locking was achieved for their laser by introducing coupling between the longitudinal modes of the fundamental laser cavity, thus permitting more efficient transmission of phase information to the edges of the laser gain bandwidth.

The dramatic pulse-shortening results obtained with nonlinear external cavity feedback have been explained by a physical and theoretical model, additive pulse mode locking (APM).[69] Table 2.5 shows recent additive mode-locking results achieved in different solid-state laser systems.[70]

## 2.5  Conclusion

There is a great deal of interest in the concept of a reliable, easy to use tunable laser for routine applications in analytical chemistry, because many laser-based techniques are available that have excellent characteristics in terms of selectivity, sensitivity, detection limits, accuracy, precision, information gathering power, and so on. Unfortunately, these techniques have not generated commercial instrumentation due to the complexity of use and maintenance of tunable lasers of various types. Dye lasers, pumped by pulsed Nd:YAG or excimer lasers, have always been capable of production of a wide range of wavelengths, especially

**Table 2.5**  Additive Pulse Mode-Locking Systems in Solid-State Lasers

| Laser | Wavelength (nm) | Performance |
|---|---|---|
| Ti:Al$_2$O$_3$ | 670–1000 | 200 fs |
| Nd:YAG | | |
|     Lamp pumped | 1060 | 6 ps |
| | 1320 | 6 ps |
|     Diode pumped | 1060 | 1.7 ps |
| Nd:YLF | | |
|     Lamp pumped | 1053 | 3.7 ps |
|     Diode pumped | 1053 | 1.5 ps |
| Nd:Glass | 1054 | 380 fs |
| NdCl:OH | 1500–1600 | 75 fs |
| KCl:Ti | 1450–1550 | 125 fs |

in conjunction with various harmonic generation techniques, but it is very difficult to routinely cover wide wavelength ranges in a single experiment. Accordingly, most scientists use dye lasers at a single wavelength or narrow range of wavelengths by use of one dye. A change in dye, especially with higher-power lasers, is done only if absolutely necessary, and with the knowledge that all experiments with the old dye should be completed before the new dye is installed. Harmonic generation of dye laser output is fairly routine and can be done under computer control, and so is much less of an issue than a dye change. The advent of lasers with wider tunable ranges than the dye laser has changed this to some extent, and these lasers will continue to erode the use of dye lasers. In particular, pulsed Ti:sapphire lasers provide a 200-nm tunable range in the red part of the spectrum that can produce some tunable radiation in the 350- to 450-nm range after harmonic generation. This is still a fairly restricted range, especially for atomic spectroscopy where most interesting wavelengths are below 350 nm. Only relatively complicated and expensive multiple stages of harmonic generation or mixing techniques can alleviate this to some extent. Pulsed Nd:YAG-pumped OPO lasers appear to hold significant promise for ultraviolet and visible atomic spectroscopy in that they produce tunable radiation under computer control from 440 to 2000 nm. The advent of BBO crystals for the optical parametric conversion process has allowed commercial versions of this type of laser to proliferate over the last few years. Problems with beam quality and pointing stability of the pump laser, usually Nd:YAG, are being addressed by several manufacturers. Certainly, an example of an OPO laser in one of the authors' laboratories (R.G.M.) has proven to be fairly reliable over several months of operating time. In addition, the use of a two-crystal harmonic generation system, also under computer control, makes this type of laser system quite easy to use over the wavelength range of 220 to 2000 nm. Unfortunately, the optical parametric oscillator, like many laser systems, still has fairly complicated optical alignment procedures after each flash laser change in the YAG pump laser. An alternative is the diode laser, but this has a rather restricted set of wavelength ranges. A different diode must be used for different discrete wavelength ranges of about 10–30 nm, and wavelengths in the blue part of the spectrum are only just becoming available for analytical spectroscopists to characterize. The devices are fairly small and inexpensive, but they require significant power supplies that cannot be described as inexpensive, especially if several of them are necessary. The electronic and mechanical engineering of combinations of diodes in one housing with integrated power supplies has not been fully prototyped and tested, and the concept of harmonic generation in these lasers has not been realized over a wide range of wavelengths. Essentially, the technology of harmonic generation would be little changed from that used for dye lasers. It would be equally expensive and complex because of the necessity for the same mechanical control of crystal angles, and the need for several crystals to generate ultraviolet radiation. Despite these engineering problems, the attraction of diode lasers lies in their electronic controllability, possible low cost when they can be mass marketed, capability for very narrow spectral linewidth, and adequate average powers for spectroscopic measurements. None of the new types of tunable lasers have been fully explored; hence they represent a potentially fruitful area of research for analytical chemists. None of these laser systems is inexpensive. A single diode laser for one 20- or 30-nm tunable range, with its power supply and without harmonic generation, starts in the $6000 to $10,000 range, while fully capable dye, Ti:sapphire, or optical parametric oscillator laser systems with harmonic generation capabilities range in price up to $200,000.

# References

1. Schawlow, A. L.; Townes, C. H., *Phys. Rev.* **1958,** *112,* 1940.

2. Maiman, T. H. *Nature* **1960,** *187,* 493.

3. *Laser Spectroscopy and its Applications;* Radziemski, L. J.; Solarz, R. W.; Paisner, J. A., Eds. Marcel Dekker: New York, 1987.

4. Andrew, D. L. *Applied Laser Spectroscopy;* VCH: New York, 1992.

5. Demtroder, W. *Laser Spectroscopy;* Springer-Verlag: Berlin, 1982.

6. *Lasers, Spectroscopy, and New Ideas;* Yen, W. M.; Levenson, M. D., Eds.; Springer-Verlag: New York, 1987.

7. *Laser Applications;* Ready, J. F.; Erf, R. K., Eds.; Academic: Orlando, 1984.

8. *Laser Applications to Optics and Spectroscopy;* Jacobs, S. F.; Sargent, M. III; Scott, J. F.; Scully, M. O., Eds.; Addison-Wesley: London, 1975.

9. *CRC Handbook of Laser Science and Technology;* Weber, M. J., Ed.; CRC: Boca Raton, FL, 1982; Vols. 1 and 2.

10. Svelto, O. *Principles of Lasers;* Plenum: New York, 1982.

11. Fox, A. G.; Li, T. *Bell Syst. Tech. J.* **1961,** *40,* 453.

12. Carswell, A.; Mankin, W. G. *Opt. Photonics News* **1993,** February, 52.

13. *Laser Focus World: Buyers Guide;* Penwell: Nashua, NH, December, 1994.

14. Brackmann, U. *Laser Dyes;* Lambda Physik: Göttingen, Germany, 1992.

15. *Dye Laser Principles with Applications;* Academic: Duarte, F. J.; Hillman, L. W. Eds; New York, 1990.

16. Schafer, F. P. *Dye Lasers;* Springer-Verlag: Berlin, 1977; Vol. 1.

17. Fork, R. L. *Opt. Lett.,* **1987,** *12,* 483.

18. Simon, P.; Szatmari, S.; Schafer, F. P. *Opt. Lett* **1991,** *16,* 1569.

19. *Dye Laser (Model LN107);* Laser Photonics: Orlando, FL, 1990.

20. *Laser Catalog;* Laser Photonics: Orlando, FL, 1990.

21. Questek, Inc., "Specifying Excimer Beam Uniformity"; Questek Technical Note 10, 1, 1991; Questek: Billerica, MA.

22. *Laser Focus World* **1993,** *June,* 135.

23. Mantz, A. W. *Microchem. J.* **1994,** *50,* 351.

24. Imasaka, T.; Ishibashi, N. *Anal. Chem.* **1990,** *62,* 363A.

25. Lawrenz, J.; Niemax, K. *Spectrochim. Acta* **1989** *44B,* 155.

26. Stern, F. In *Laser Handbook; Arecchi, F. T.; Schulz-Dubois, E.O. Eds.; Northe Holland: Amsterdam,* **1972** Chapter B4.

27. Kressel, H.; Butler J. K.; *Semiconductor Lasers and Heterrojuction LEDs;* Academic: New York, **1977.**

28. Thompson G. H.B.; *Physics of Semiconductor Laser Devices; Wiley: New York,* **1980.**

29. Hergenroder, R.; Niemax, K. *Trends Anal. Chem.* **1989,** *8,* 333.

30. Hergenroder, R.; Niemax, K. *Spectrochim. Acta* **1988,** *43B,* 1443.

31. Groll, H.; Schaldach, G.; Berndt, H.; Niemax, K. *Spectrochim. Acta* **1995,** *50B,* 1293.

32. Ng, K. C.; Ali, A. H.; Barrer, T. E.; Winefordner, J. D. *Appl. Spectrosc.* **1990,** *44,* 849.

33. Snitzer, E. *Phys Rev. Lett.* **1961,** *7,* 444.

34. Etzel, H. W.; Gandy, H. W.; Ginther, R. J. *Appl. Opt.* **1962,** *1,* 534.

35. Snitzer, E.; Woodcock, R. *Appl. Phys. Lett.* **1965,** *6,* 45.

36. Gandy, H. W.; Ginther, R. J.; Weller, J. F. *J. Appl. Phys.* **1967,** *38,* 3030.

37. Gandy, H. W.; Ginther, R. J. *Proc. IRE* **1962,** *50,* 2113.

38. Andreev, S. I.; Bedilow, M. R.; Karapetyan, G. O.; Likhachev, V. M. *Sov. J. Opt. Tech.* **1967,** *34,* 819.

39. Johoson, L. F.; Dietz, R. E.; Guggenheim, H. J. *Phys. Rev. Lett.* **1963,** *11,* 318.

40. Mollenauer, L. F.; Olson, D. H. *J. Appl. Phys.* **1975,** *46,* 3109.

41. Walling, J. C.; Jenssen, H. P.; Morris, R. C.; O'Dell, E. W.; Peterson, O. G. *Opt. Lett.* **1979,** *4,* 182.

42. Moulton, P. F. *J. Opt. Soc. Am.* **1986,** *B, 3,* 125.

43. Giordmaine, J. A.; Miller, R. C. *Phys. Rev. Lett.* **1965,** *14,* 973.

44. Ebrahhimzadeh, M.; Dunn, M. H. *Opt. Commun.* **1988,** *69,* 161.

45. Henderson, A. J.; Ebrahhimzadeh, M.; Dunn, M. H. *J. Opt. Soc. Am.* **1990,** *B, 7,* 1420.

46. Tang, C. L.; Bosenberg, W. R.; Ukachi, T.; Lane, R. J.; Cheng, L. K. *Laser Focus World* **1990,** October, 107.

47. Higgins, T. V. *Laser Focus World* **1994,** August, 67.

48. Simon, U.; Tittel, F. K. *Laser Focus World* **1994,** May, 99.

49. Ebrahhimzadeh, M.; Henderson, A. J.; Dunn, M. H. *IEEE J. Quantum Electron.* **1990,** *26,* 1241.

50. Orr, B. J.; Johnson, M. J.; Haub, J. G. In *Tunable Laser Applications;* Duarte, F. J., Ed.; Marcel Dekker: New York, 1995, Chapter 2.

51. Johnson, B. C.; Newell, V. J.; Clark, J. B.; Mcphee, E. S. *J. Opt. Soc. Am.* **1995,** *B, 12,* 2122.

52. *Quanta-Ray MOPO-730 Optical Parametric Oscillator Instruction Manual;* Spectra-Physics: Mountain View, CA, 1994.

53. Franken, P. A.; Hill, A. E.; Peters, C. W.; Weinreich, G. *Phys. Rev. Lett.* **1961,** *7,* 118.

54. Wallenstein, R. In *Frontiers of Laser Spectroscopy of Gases;* Alves, A. C. P.; Brown, J. M.; Hollas, J. M. Eds.; Kluwer Academic: Dordrecht, 1988, p 53.

55. Stoicheff, B. P. In *Frontiers of Laser Spectroscopy of Gases;* Alves, A. C. P.; Brown, J. M.; Hollas, J. M. Eds.; Kluwer Academic: Dordrecht, 1988, p 63.

56. Andrews, D. L. *Lasers in Chemistry;* Spring-Verlag: New York, 1986.

57. Shen, Y. R. *The Principle of Nonlinear Optics;* Wiley: New York, 1984.

58. Fowles, G. R. *Introduction to Modern Optics;* Reinhat and Winston: New York, 1975.

59. *Advances in Laser Spectroscopy;* Garetz, B. A.; Lombardi, J. R. Eds.; Wiley: New York, 1986, Vol. 3.

60. Funayama, M.; Mukaihara, K.; Morita, H.; Okada, T.; Tomonaga, N.; Izumi, J.; Maeda, M. *Opt. Comm.* **1993,** *102,* 457.

61. Wirth, M. J. *Anal. Chem.* **1990,** *62,* 270A.

62. New, G. H. C. *Rev. Prog. Phys.* **1983,** *46,* 877.

63. French, P. M. W.; Gomes, A. S. L.; Gouveia-Neto, A. S.; Taylor, J. R. *Opt. Comm.* **1986,** *60,* 389.

64. MacFarlane, D. L.; Casperson, L. W.; Tovar, A. A. *J. Opt. Soc. Am.* **1988,** *B, 5,* 1134.

65. Mollenauer, L. F.; Stolen, R. H. *Opt. Lett.* **1984,** *9,* 13.

66. Migschke, F. M.; Mollenauer, L. F. *Opt. Lett.* **1987,** *12,* 407.

67. Blow, K. J.; Wood, D. *J. Opt. Soc. Am.* **1988,** *B, 5,* 625.

68. Blow, K. J.; Nelson, B. P. *Opt. Lett.* **1988,** *13,* 1026.

69. Ippen, E. P.; Haus, H. A. *J. Opt. Soc. Am.* **1989,** *B, 6,* 1736.

70. Fujumoto, J. G.; *Opt. Photonics News* **1991,** *3,* 9.

CHAPTER

# 3

# Laser-Excited Atomic Fluorescence Spectrometry: Principles, Instrumentation, and Applications

*Xiandeng Hou, Suh-Jen Jane Tsai,*
*Jack X. Zhou, Karl X. Yang,*
*Robert F. Lonardo, and Robert G. Michel*

## 3.1 Introduction

Laser-excited atomic fluorescence spectrometry (LEAFS) involves the excitation of analyte atoms by a laser and subsequent detection of fluorescence. By use of a laser as the light source, the population of excited analyte atoms is much higher than with a conventional light source. Thus, LEAFS is extremely sensitive, with detection limits as low as a few femtograms. The merits of LEAFS also include calibration curves with wide linear dynamic range and excellent spectral selectivity. The instrumentation of LEAFS is composed of the laser, a device for production of gas-phase atoms, and a fluorescence detection system. LEAFS has been successfully used for both analytical and diagnostic purposes.

In 1971, seven years after atomic fluorescence spectrometry (AFS) was demonstrated as a potential analytical method by Winefordner and his coworkers,[1,2] Denton and Malmstadt[3] used a commercial, frequency-doubled, Q-switched ruby laser to pump a home-built dye laser to excite atomic fluorescence in a flame, while Fraser and Winefordner[4] employed a commercial nitrogen-pumped dye laser as the excitation source for flame LEAFS. In the same year, Kuhl and Marowsky[5] employed a laboratory-constructed flashlamp-pumped dye laser, also for flame LEAFS. It was shown that laser excitation gave significant improvements in limits of detection (LOD) and linear dynamic ranges (LDR) of calibration curves compared to conventional source excitation. Since then, LEAFS has been advanced by use of improved dye laser technologies.

This chapter deals with the principles, instrumentation, analytical applications, and future prospects of laser-excited atomic fluorescence spectrometry. Conventional source–excited atomic fluorescence spectrometry is not included here, for which the reader is referred to an article by Butcher et al.[6] The use of LEAFS for diagnostic purposes is not covered in this chapter, either. Articles by Omenetto et al.[7] and Hill et al.[8] should be consulted for this important aspect of LEAFS.

## 3.2 Theory

Detailed discussion on the basic theories and mathematical expressions relevant to atomic fluorescence with laser excitation can be found in many articles.[7,9–24] In this section, the significance of optical saturation, factors that affect the calibration curves, sources of noise, and detection limit are discussed from a practical point of view. A discussion of selectivity and precision is also included.

### 3.2.1 Optical Saturation

When a light source is applied to a ground-state atomic vapor, only the valence, or outermost, electrons are excited from the ground state to the excited state. The excited atoms then undergo chemical and physical processes, of which atomic fluorescence is one of the physical processes involved besides inelastic collisions and stimulated emission. Equation (3.1) shows a general expression for fluorescence intensity, with symbols defined below.

$$B_F = \ell/(4\pi) \, h\nu_{ul} A_{ul} n_u \tag{3.1}$$

For a two-level atom and low excitation irradiance, the expression becomes:

$$B_F = \ell/(4\pi) \, Y_{ul} E_v(\nu_{ul}) \int k(\nu) \, d\nu \tag{3.2}$$

$$Y_{ul} = A;_{ul}/(k_{ul} + A_{ul}) \tag{3.3}$$

When high excitation irradiance is applied, this becomes:

$$B_F = \ell/(4\pi) h\nu_{ul} A_{ul} n_T g_u/(g_u + g_l) \tag{3.4}$$

For Equations (3.1) through (3.4), the symbols are defined as follows:

$B_F$ = the atomic fluorescence radiance, J s$^{-1}$ cm$^{-2}$ sr$^{-1}$
$h\nu_{ul}$ = energy of emission photon, J
$\nu_{ul}$ = frequency of emission photon, s$^{-1}$
$n_u$ = upper (radiative) level population density, m$^{-3}$
$n_T$ = total population density of all electronic states of same atom, m$^{-3}$
$g_l, g_u$ = statistical weights of lower and upper levels, respectively

$\ell$      = emission fluorescence path length, m
$Y_{ul}$    = the quantum efficiency of the transition $1 \rightarrow 2$
$E_v(v_{ul})$ = source spectral irradiance at $v_{ul}$, J s$^{-1}$ m$^{-2}$ Hz$^{-1}$
$k(v)$      = atomic absorption coefficient at frequency v, m$^{-1}$
$k$         = pseudo-first-order radiationless rate coefficient between the transition levels, s$^{-1}$
$A$         = Einstein coefficient of spontaneous emission between the transition levels, s$^{-1}$

At low source intensity, the deexcitation rate is faster than the excitation rate. Under such circumstances, the fluorescence signal $B_F$ is dependent on the product of the source intensity $E_v$ and the quantum efficiency $Y_{21}$ as shown in Equation (3.2). At high source intensity, the rate of the deexcitation is close to the rate of the excitation, or stimulated absorption (about $10^8$–$10^9$ s$^{-1}$). In this case, the fluorescence signal is independent of the light intensity, as expressed in Equation (3.4).

If the fluorescence signal is plotted versus source intensity, a saturation curve can be obtained, as shown in idealized form in Figure 3.1.[20,25] The saturation curve has a slope of one at low source intensities, the linear region, and a slope of zero at high source intensities, the saturation region.

At low source intensities, the rate of excitation by the source is much less than the rate of deexcitation, and therefore, the majority of the atoms remain in the ground state. The fluorescence signal is directly proportional to both the

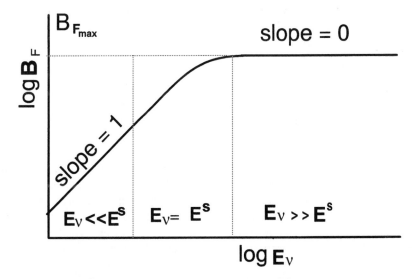

**Figure 3.1**   An idealized saturation curve: fluorescence signal log $B_F$ versus source energy log $E_v$. $E^s$ is the saturation energy of the transition. Reproduced with permission from reference 20.

source intensity and the quantum efficiency, because the steady-state population of the excited state is, essentially, governed by the quenching rate coefficient. Thus, the maximum possible fluorescence signal is not obtained under these conditions.

Optical saturation can be obtained by use of a laser for which the source-induced radiative excitation and deexcitation rate coefficients dominate the mechanisms of population of the energy levels. Thus, the fluorescence signal reaches a plateau to become independent of the source intensity and the quantum efficiency. However, further enhancement of the laser irradiance will result in increased nonspecific scatter signals, such as concomitant scatter and stray light, which are directly proportional to the source intensity even at high spectral irradiance.[26,27] The minimum laser irradiance required to saturate the transition should be employed to minimize such scatter interference.

The use of a high energy laser can also result in the reduction of ground-state analyte population through saturation broadening.[7,9,22,23,28-31] When a high-energy narrowband laser saturates a population of atomic transitions, fewer atoms are in the ground state than if the laser were not present. This reduces the absorption coefficient at the atomic spectral line center and limits the fluorescence signal there. Conversely, the absorption coefficient at the wings of the line is less affected by the intense laser beam[23] such that, with increased laser power, relatively more absorption occurs at the wings of the line compared to the center. The result is saturation-broadened absorption and fluorescence lines.

Several types of fluorescence transitions have been observed with laser excitation, as shown in Figure 3.2, including resonance fluorescence, Stokes direct line fluorescence, anti-Stokes direct line fluorescence, stepwise fluorescence, thermally assisted fluorescence, double-resonance excitation fluorescence, and two-step excitation fluorescence.

### 3.2.2 Calibration Curves

Like almost all other optical spectroscopic methods, the LEAFS measurement is a relative analytical method, and its accuracy is largely dependent on the reliability of the calibration curve, which is also called the fluorescence growth curve.

The linear dynamic range (LDR) of the calibration curve of LEAFS is excellent because it extends to six or seven orders of magnitude. This is a result of the very high sensitivity, which extends the LDR down to very low concentrations, and the ability to saturate the atoms with the laser, which extends the LDR to high concentrations. This is a very significant advantage, compared with atomic absorption in a graphite furnace or a flame where the LDR is only a couple of orders of magnitude. An improved LDR makes multiple dilution of sample solutions less necessary.

Theoretically, fluorescence intensity is directly proportional to the analyte concentration. In practice, however, the shape of the calibration curve is dependent on the experimental parameters.[32-35] At low concentrations, the cali-

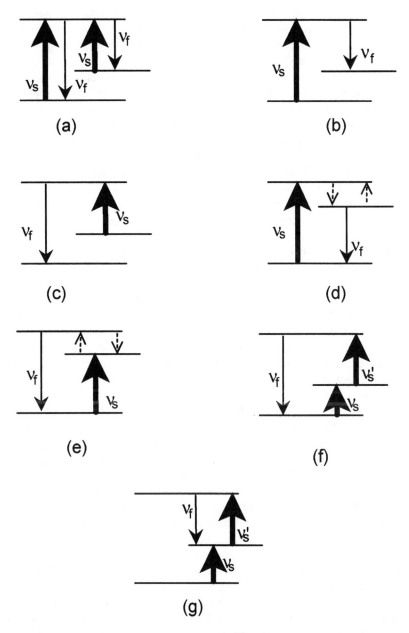

**Figure 3.2** Several types of fluorescence transitions observed with laser excitation. $v_s$-excitation frequency; $v_f$-fluorescence frequency. (a) resonance fluorescence; (b) Stokes direct line fluorescence; (c) anti-Stokes direct line fluorescence; (d) stepwise fluorescence; (e) thermally assisted fluorescence; (f) double-resonance excitation fluorescence; (g) two-step excitation fluorescence.

▲ absorption transition; ↑ fluorescence transition; ↕ thermal transitions.

bration curve is a linear function of the analyte concentration irrespective of the type of laser source. At high concentrations, the calibration curve will have a slope of zero with broadband continuum laser excitation, and a negative slope in the case of a narrowband laser excitation source, where a rollover in the calibration curve can be observed. In practice, rollover is seldom observed for excitation with common pulsed dye lasers. In addition, there is a strong relationship between the shape of the calibration curve and the degree of saturation. Without saturation, the linear dynamic range of the calibration curve is worsened due to self-absorption at high concentrations.[28,36]

Self-absorption is the reduction of the fluorescence signal due to the reabsorption of atomic fluorescence by analyte atoms in the irradiated portion of the atom cell. If the measurement of the analyte signal is done under optical saturation, the high degree of excitation leaves few atoms to cause self-absorption, and the vapor becomes transparent to the resonance line radiation. The reduction of self-absorption results in an improvement of about two orders of magnitude in the LDR of LEAFS compared with conventional source–excited AFS.[10] Weeks et al.[37] showed the improvement of the calibration curve experimentally.

Other factors of importance include[25,38] the prefilter effect and postfilter effect, which are dependent on the fluorescence detection mode. For the right-angle detection mode, there may be both a prefilter and a postfilter effect. However, there exists neither a prefilter effect nor a postfilter effect in the front-surface detection mode,[39-45] in which fluorescence is collected in a direction 180° to the laser beam axis. Figure 3.3 shows the prefilter and postfilter effects schematically.

In the case of the prefilter effect for right-angle detection at high analyte concentrations, the fluorescence intensity may be reduced by the absorption of source light by atoms that are near the front of the illuminated volume, where fluorescence is not detected (Figure 3.3c–d). This phenomenon causes bending of the calibration graphs. The prefilter effect is not physically possible with front-surface detection because all of the fluorescence from atoms near the front of the illuminated volume is detected, as shown in Figure 3.3e.

The postfilter effect is a special case of self-absorption that applies only to the right-angle detection mode. This is the reduction in fluorescence intensity at high analyte concentrations that occurs within the detection volume, but outside the illuminated volume, as shown in Figure 3.3b and d. The postfilter effect is not physically possible for front-surface detection because the atom cell is completely irradiated in the direction of the detector. The postfilter effect may cause significant reduction of the LDR of LEAFS even under saturation conditions for right-angle detection. Furthermore, the postfilter effect occurs in both resonance and nonresonance transitions even when the atomic transition is saturated. The possibility for postfilter and self-absorption effects to occur is higher for those transitions that terminate in the ground state or in a relatively long-lived excited state, or metastable state, than for those transitions that terminate in a short-lived excited state.[35]

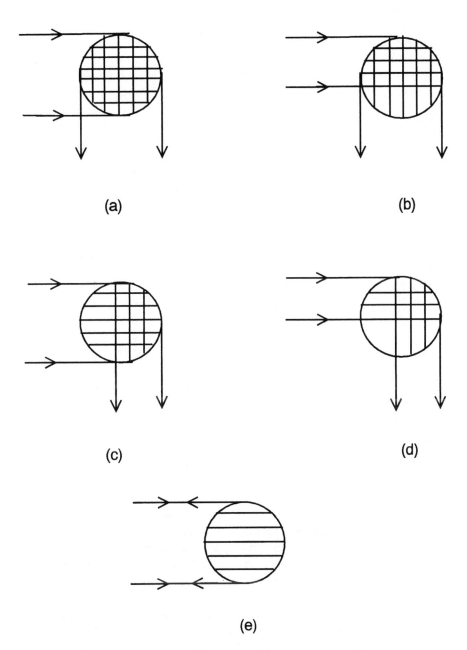

**Figure 3.3** Schematic diagram of different illumination and observation geometries of right-angle (*a–d*) and front-surface (*e*) fluorescence showing a top view of the cell. (*a, e*) No prefilter and postfilter effects; (*b, d*) postfilter effects; (*c, d*) prefilter effects.

The influence of the postfilter effect upon the LDR for resonance fluorescence of thallium in a flame was studied by Omenetto et al.[36] When the atoms on the edge of the flame near the detector were irradiated, the postfilter effect was not present, and the calibration curve remained linear up to $10^4$ ppm. However, if the edge of the flame away from the detector was illuminated, a severe postfilter effect occurred, which reduced the LDR by an order of magnitude. If both the pre- and postfilter effects were present, then the bending of the calibration curve occurred at even lower concentrations and caused further reduction in LDR.

Although the self-absorption or postfilter effect can be eliminated or reduced by nonresonance conventional source AFS, self-absorption may still be observed if a laser is employed.[7] With nonresonance conventional source AFS, there is no self-absorption or postfilter effect for a Stokes direct line transition, as shown in Figure 3.2, because the population of the intermediate level is small, with the result that there are few atoms capable of absorption of the nonresonance fluorescence. When a laser light source is used, the situation can be different. For example, in the Stokes direct line fluorescence transition of lead, the excitation wavelength is 283.3 nm and fluorescence is detected at 405.8 nm. The intermediate level is metastable with a lifetime of 10 μs. Bolshov et al.[46,47] have shown that 40% of the lead atoms are transferred into this level during a 5-ns pulse that saturates the transition. Self-absorption can occur for such a nonresonance transition because a significant population occupies the intermediate level and can absorb the nonresonance fluorescence. The effects of metastable intermediate levels on the calibration curve for lead and copper have been observed by Dougherty et al.[35] Their work also shows a high probability for reabsorption of fluorescence of nonresonance LEAFS of silver because it consists of a Stokes stepwise transition.

### 3.2.3 Sources of Noise and Limit of Detection

#### 3.2.3.1 Sources of Noise

The sources of noise in LEAFS have been divided into five major categories by Piepmeier.[24] These are (1) background fluorescence shot and flicker noise from molecular fluorescence and nonanalyte atomic fluorescence; (2) background radiation shot and flicker noise, such as atom cell emission; (3) source scatter shot and flicker noise, such as concomitant scatter and stray light; (4) analyte fluorescence and emission shot and a flicker noise; and (5) interferent fluorescence and emission shot and flicker noise. In addition, radio frequency interference and the detection electronics can be considered as sources of noise.[25]

The scattering of laser radiation off particles introduced by the sample is defined as concomitant scatter.[48] Concomitant scatter is expected to be the most significant for the analysis of involatile and undissociated matrices. Its magnitude can be minimized by use of a laser with spectral linewidth equal to or less than that of the absorption linewidth.[49] This means that there will be very little unabsorbable radiation in the laser light and, hence, the minimum possible light available to be scattered.

In a flame, concomitant scatter may be significant for real sample analyses because the chemical environment of the atom cell may generate nonvolatile species of high scattering potential, such as oxides.[7] When the analyte is contained in an inert argon atmosphere (as in a furnace), concomitant scatter is reduced compared to when it is in a flame. Concomitant scatter is serious in cup furnace atomization, because fluorescence is detected in a relatively cool region above the cup. In this cool region, recombination of matrix elements or the analyte may form scattering species.[50] Less recombination and the associated scatter are expected in tube furnaces because the volume inside the tube is maintained at a relatively high temperature.[51] A tube furnace coupled with front-surface illumination and detection does not require ports for passage of the laser beam.[39,40] The tube furnace has been shown to reduce vapor-phase interference greatly in atomic absorption, compared with open furnaces (e.g., cups).[51] However, with a highly involatile scattering matrix, concomitant scatter may still be the limiting source of noise for an enclosed tube furnace. Some detailed discussion can be found in the literature.[52,53]

Unlike scattered radiation, stray light is defined as the laser light reflected into the monochromator from various parts of the instrument, but not scattered from the sample itself. Like concomitant scatter, stray light can be minimized by use of a narrow spectral band laser. Stray light is usually the limiting noise in resonance LEAFS for flames, plasmas, and furnaces. However, stray light may also be significant in nonresonance LEAFS because monochromators with wide bandpass (1–4 nm) and poor stray light rejection are commonly used to detect fluorescence. Extensive discussion is found in references 39, 48, 49, 52, and 54–56.

Atom cell background emission is also a source of noise. In most cases, atom cell emission increases with the detection wavelength. Flame background emission is dependent on the flame type (usually air-acetylene or dinitrogen oxide–acetylene) and the temperature of the flame.[48,56] For electrothermal atomization LEAFS (ETA-LEAFS), the magnitude of blackbody emission from the atomizer is dependent on the type of atomizer (cup, tube, or rod), the detection wavelength, and the atomization temperature. At high atomization temperatures and long detection wavelengths, ETA blackbody emission can degrade the limit of detection. Furnace emission must be minimized in tube furnace LEAFS by baffling radiation from the walls of the tube and imaging the center of the tube on to the entrance slit of the monochromator.[25,45,52] The inductively coupled plasma (ICP) has a high spectral background that is significant enough to have reduced the utility of ICP-LEAFS.[57–59]

Molecular fluorescence may be generated by species commonly present in air-acetylene and dinitrogen oxide–acetylene flames, such as OH, CN, $C_2$, metal chlorides, metal oxides, and metal hydroxides.[60–64] Molecular fluorescence may be a limiting noise source if it overlaps the atomic fluorescence line.

With a graphite furnace ETA, molecular fluorescence noise must originate primarily from the sample, because atomization occurs in an inert argon atmosphere. Many molecules have been detected by furnace atomic absorption,[65–67] and it is possible that these molecules may fluoresce. In the cup furnace, atomization occurs in a relatively cool environment above the cup where significant molecule formation may be expected.[19] The tube furnace instrument that employs right-angle detection has laser ports in the walls of the furnace that may allow cooling of the interior of the furnace, which, in turn, may also allow some condensation and subsequent molecule formation. The tube furnace that uses front-surface detection has no laser ports in the walls of the tube. As a result, atomization occurs into a relatively hot environment, which should minimize molecule formation.[65–67] Therefore, molecular fluorescence noise, if it is present, should be minimized by front-surface detection in a tube furnace.

Nonanalyte atomic fluorescence in LEAFS is seldom a difficult problem to solve. For example, Weeks et al.[37] reported a spectral interference for resonance LEAFS of manganese in the presence of gallium. Both elements were excited at 403.3 nm in an air-acetylene flame by means of a dye laser with a 0.01 nm bandwidth. This interference was eliminated by excitation of manganese at the 403.1 nm resonance line. Spectral overlaps can be minimized in all atom cells by using a laser with a linewidth equal to or less than that of the atomic linewidth (approximately 0.002 nm).

Significant radio frequency interference can be generated by certain high-power, high repetition rate, pulsed laser systems (e.g., excimer, Nd:YAG, and copper vapor lasers). Rutledge et al.[68] reported that radio frequency was a significant noise source for their instrument that employed a copper vapor–pumped dye laser. With most laser systems this radio frequency noise can be minimized by proper shielding of the pump laser and the detection electronics.

The limiting noise produced by the boxcar or the photomultiplier tube is designated as electronic noise. Epstein et al.[59] reported that electronic noise is a significant noise source at low radio frequency powers for ICP-LEAFS. At higher plasma powers, electronic noise did not seem to affect the signal-to-noise ratio significantly. Electronic detector noise was only 3–5% of the laser scatter or flame emission noise with the 0.5 mm slit width used in flame LEAFS work performed by Walton et al.[42]

### 3.2.3.2 Limit of Detection

The limit of detection (LOD)[10,19,54] is usually defined as the amount of analyte in a sample that produces a signal equivalent to three times the standard deviation of the noise associated with the blank. Long and Winefordner[69] discussed the LOD, and pointed out that it is affected by the reliability of the calibration curve. The LOD can be found by relating the equivalent signal mentioned above to a concentration or mass, by dividing by the slope of the calibration curve obtained from a linear regression analysis. For steady-state atomizers (flames and plasmas), the LOD is generally expressed in concentration units ($c_a$). Mass LODs are commonly used for non-steady-state atomizers, such as graphite electrothermal atomizers and some vapor cells, to represent the mass of analyte in the cell ($m_a$).[10] Although concentration LODs have also been employed to define the LOD for non-steady-state atom cells, their use is ambiguous because different sample volumes of analyte solution may be injected. The same instrument may have concentration LODs that vary with the sample volume.

The LOD is dependent on laser type, the atomic transitions, the atom cell, and the fluorescence detection arrangement. In addition, maintenance of saturation of the relevant energy levels is essential to achieve the best LOD. For graphite furnace LEAFS, the LOD can be as low as a few femtograms.[38,52,54,70–74] In order to obtain the intrinsic LOD, for which the main limiting noise is not due to the background or the detector but the inherent statistical fluctuation in the number of atoms present in the volume probed by the laser, a burst of atoms must be produced in a time equal to the probe time, which is equivalent to the duration of the laser pulse.[7,16,19] The probe volume must be as close to the atomizer volume as possible. This can be achieved by optimization of the laser beam size at the atomizer under saturated conditions. Different detection systems give different LODs. For example, as in furnace LEAFS,[75] the experimental setup needs to be optimized to allow the minimum noise due to furnace emission, and rejection of stray laser photons in the detection system. This can be achieved by measurement of the fluorescence in the ultraviolet, where the emission from the hot furnace is negligible. The detection system must give maximum collection efficiency of the fluorescence light. Detailed discussion on the overall efficiency of LEAFS is given in references 25 and 45.

### 3.2.4 Selectivity

Besides high sensitivity, high selectivity is the other main feature of LEAFS. Laser excitation provides more selective access to nonresonance transitions than a conventional light source. Thus, the possibility of spectral interference is reduced. The use of a high-power laser beam will result in the broadening of the fluorescence excitation profiles, by phenomena such as saturation broadening mentioned in section 3.2.1, and the spectral selectivity may be degraded to some degree.[72] However, compared with ICP-AES, the fluorescence technique suffers considerably less spectral interference,[75] such that it is possible to do direct analysis of rare earth elements by ICP-LEAFS without prior chemical separation. This is due to the high excitation selectivity based on single-line laser excitation, and the selectivity of the fluorescence wavelength. The atomic fluorescence spectrum of the analyte is also much simpler than its atomic emission spectrum because multiple excitation is minimized.[75]

### 3.2.5 Precision

Generally, the precision of LEAFS is slightly poorer than that obtained by conventional AFS, AAS, and ICP-AES. This may be due to the pulse-to-pulse stability of pulsed dye lasers. The precision is approximately 3–5% under optimum conditions. Usually, the precision of ETA-LEAFS is similar to graphite furnace atomic absorption work, plus a possible contribution from laser pulse-to-pulse fluctuation. Also, the precision improves with the square root of the repetition rate of the laser.[56,76–78]

## 3.3 Instrumentation

In this section, the instrumentation of LEAFS (Figure 3.4) is discussed, including lasers, atom cells, optics, detection systems, and background correction techniques. In addition, isotope determinations by ICP-LEAFS and Doppler-free spectroscopic methods conclude this section.

### 3.3.1 Lasers

The lasers used in LEAFS must be wavelength tunable, and capable of generating fairly high peak energy and high average power to probe atoms with sensitive lines almost anywhere in the visible and ultraviolet spectrum. This allows the determination of as many elements as possible.[25,76,77]

Many kinds of tunable laser can be used in LEAFS, although the dye laser has been regarded as the only tunable laser for LEAFS. Transition metal ion vibronic-state lasers[79] and titanium:sapphire lasers[80] have potential for LEAFS.

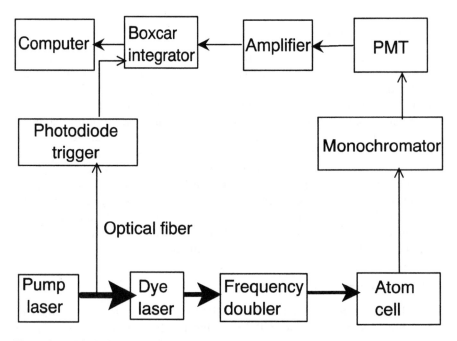

**Figure 3.4**   Block diagram of LEAFS instrumentation.

In addition, optical parametric oscillator[80,81,82] and tunable semiconductor diode lasers[83] have possible usefulness in LEAFS.

The pump lasers that have been used in LEAFS include frequency-doubled ruby lasers,[3] flashlamp-pumped dye lasers,[30] Nd:YAG lasers,[46,47,49,50,71,84] nitrogen lasers,[4,37,85-87] argon ion lasers,[88] copper vapor lasers (CVLs),[70,89-92] and excimer lasers.[52,70,73,77,93,94] The most popular pump laser in LEAFS is the excimer laser,[12,42-44,57,95] and the argon ion laser is seemingly the only CW laser that has been used in LEAFS. Most LEAFS research[4,28,37,41,46,47,49,71,75,85,93,95-97] has been carried out with repetition rates of 50 Hz or less, although high repetition rate (800–10,000 Hz) copper vapor–pumped dye lasers have been used for LEAFS in graphite furnaces[70,91] and the pulsed glow discharge,[92] and a 500-Hz excimer laser has been used for ETA-LEAFS.[43] In general for flames and plasmas, a repetition rate of 30 Hz is good enough to obtain adequate measurement precision. Compared with flame LEAFS, higher repetition rates are required for ETA-LEAFS because the fluorescence signals in the electrothermal atomizer are transient and need to be sampled at a greater rate to improve the limit of detection.

The most important region of the spectrum for LEAFS, between 195 and 320 nm, can not be directly obtained with either dye lasers or any other type of laser. However, it can be obtained through optical frequency conversion techniques,[98-107] frequency doubling, frequency mixing, and stimulated Raman scattering (SRS). For example, Leong et al.[98] evaluated SRS of tunable dye laser radiation as a primary excitation source for ICP-LEAFS. In SRS, the shifted wavelengths are coherent and as monochromatic as the initial pump laser system. The tunability of the scattered Raman radiation is determined by the tunability of the laser employed. For a dye laser, with its limited tuning range, the scattered Raman radiation does not cover the entire ultraviolet-visible light wavelength range, but only discrete regions of it, with gaps between the anti-Stokes wavelength regions, as shown in Figure 3.5. Pump power, the pressure of the gas inside the SRS cell, and absolute temperature of the scattering gas are adjusted to achieve the maximum radiant flux of a shifted wavelength.

The advent of the β-barium borate crystal has now allowed frequency doubling and mixing down to about 195 nm.[100] For more detailed information on principles of lasers and frequency conversion techniques, readers are encouraged to consult Chapter 2.

### 3.3.2 Atom Cells

Flames have been very popular atom cells for AAS because of their simplicity and reliability.[25] For this reason, most early work on LEAFS involved the use of flames.[4,7,34,56,85,88,94,95,108-116] Plasmas[57-59,75,117-119] and glow discharge devices[120,121] have also been used. Extremely low intrinsic LODs can be achieved by ETA-LEAFS[13,39-41,46,47,49,50,52,71,73,74,77,122-128] and glow discharge devices.[100,120,121]

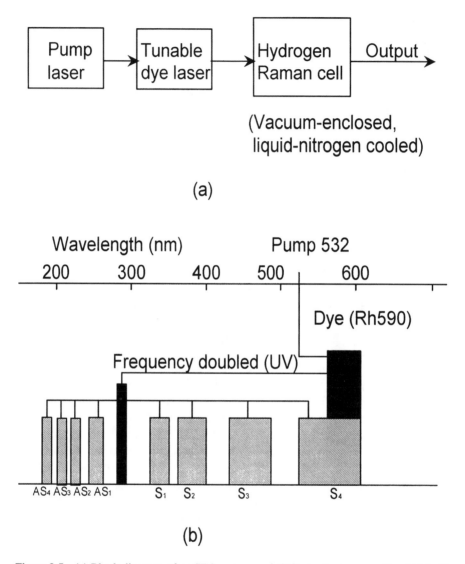

**Figure 3.5** (*a*) Block diagram of an SRS system and (*b*) its tuning ranges of input (*dark*) and output (*cross-hatched*) radiations. Modified (*a*) and reproduced (*b*) with permission from reference 98.

### 3.3.2.1 Flames

In flame LEAFS, the signal-to-noise ratio of a burner with a long path length is better than that of a burner with a shorter one. In the determination of manganese by flame LEAFS, Walton et al.[42] found that front-surface detection was more efficient than transverse, right-angle detection, and that the signal size observed for a long path-length flame (100 mm) was about fourfold larger than that observed for a shorter flame (10 mm). Theoretically, a tenfold signal increase should be observed between the two flames because of the proportionality between absorption and the path length of the atom cell. The fact that only a fourfold increase was observed suggested that the front-surface mirror arrangement was unable to collect all of the fluorescence radiation emitted from the full length of the long path-length flame. It was also observed that the total noise measured in the long path-length flame was approximately twice that measured in the shorter one. The increase in noise was most likely due to greater scatter of laser radiation in the long path-length flame compared with that from the shorter path-length flame. Overall, the LOD for manganese for the long path-length flame LEAFS was approximately three times better than that of the shorter one.

For LEAFS, the flame may be surrounded by sheath of inert gas (i.e., nitrogen or argon) to minimize the interaction of the flame gases with air.[116] This decreases the flame background and reduces quenching of the fluorescence signals.[129,130] Noise correlation studies on the simultaneous detection of laser-enhanced ionization and laser-induced fluorescence in flames have been done by Turk and Travis.[131]

In our laboratory,[42] flame LEAFS has been shown to be a high-sensitivity detector for organomanganese and organotin compounds. In connection with high performance liquid chromatography (HPLC), the signal and noise characteristics of instrumentation for dispersive, nondispersive, and front-surface LEAFS were studied.

With a hydrogen-air flame and/or an argon ICP, Hueber et al.[132] determined arsenic by argon fluoride LEAFS. A simple fixed-frequency ArF excimer laser was used to produce a broadband output centered at 193.0–193.2 nm that overlapped the arsenic absorption line at 193.7 nm. Detection limits of about 20 ng/mL were obtained with both atomizers, and laser-induced scatter was the limiting noise in both cases.

Butcher et al.[25] summarized the detection limits obtained by flame LEAFS, flame AAS, and ETA-AAS. Virtually no recent work has been done with flame LEAFS, although improvement in technology in recent years should result in improvement of the detection limits.

### 3.3.2.2 Plasmas

The ICP provides several advantages compared to a flame. Atoms and ions can be effectively excited to about 2 eV above the ground state,[24] and the ICP has high quantum efficiency for many elements. ICP–optical emission spectrometry (ICP-OES) has relatively poor selectivity due to spectral interference and background emission, although better selectivity can be obtained when laser-excited fluorescence is detected instead of emission. This is because the resolution is determined by the laser linewidth in LEAFS rather than by the detection system.

The atomic fluorescence of silver, gold, hafnium, iridium, molybdenum, niobium, palladium, platinum, ruthenium, tantalum, and zirconium has been studied in the ICP. The calibration graphs obtained had a linear dynamic range of over four orders of magnitude, except for Au for which only three orders of magnitude were obtained.[133] LODs for a number of rare earth elements have been determined by nonresonance fluorescence spectrometry and double-resonance flourescence (DRF).[134,135] A preliminary study of a typical spectral interference of two rare earth elements, dysprosium and gadolinium, showed that the high spectral selectivity of one-step laser-induced fluorescence can be degraded at high laser intensities.[136] A standard commercial ICP unit was used in the first work on DRF by Omenetto et al.[137] The selectivity was increased by use of double-resonance ICP-LEAFS. The LODs obtained for a number of elements by ICP-LEAFS are comparable to those achieved by ICP-AES.[98] Readers are referred to a comprehensive review on inductively coupled plasmas in atomic fluorescence spectrometry for more detailed discussion on the ICP cell in LEAFS.[138]

Uranium isotopic abundance has been determined by ICP-LEAFS[139] with excitation and fluorescence wavelengths of 286.57 and 288.96 nm, respectively. The limit of detection obtained was about 2 µg/mL. The nonresonance fluorescence determination of uranium was free from spectral line interference in a complex matrix.

Butcher et al.[25] compared detection limits obtained with ICP-LEAFS, ICP–laser-enhanced ionization fluorescence spectrometry, ICP-OES and ICP-MS. For most of the elements, the ICP-MS detection limits were the best. Within a factor of five, the fluorescence detection limits were better than ICP-OES for 2 elements, the same for 7 elements, and worse for 12 elements.

A limited amount of work has been carried out using a three-electrode direct current plasma (DCP) as an atom cell for LEAFS.[119] Limited data indicated that detection limits obtained with the DCP and ICP are not significantly different.

An atomizer that was a combination of a microwave-induced plasma in atmospheric helium and a filament vaporization system has demonstrated sensitive detection of trace elements in pure water.[87] Sixteen elements have been analyzed by this system. The detection limit was less than 1 ng/mL for most of these elements, which was better than the results with flame LEAFS.

### 3.3.2.3 Electrothermal Atomizers

Electrothermal atomizers are now the most popular atomizers in LEAFS because of their ability to obtain subfemtogram detection limits.[72] The progress achieved has been in electrothermal atomizer designs, optical arrangements for collection of fluorescence, and various demonstrations of analyses of real samples. Generally, all of the studies on ETA-LEAFS have been done with pulsed laser excitation, and mainly with XeCl excimer–pumped dye lasers.[140,141] Both the peak height and peak area of the transient signal have been measured, analogous to ETA-AAS, with the precision in the range of 5–10%.

Several electrothermal atomizer designs have been evaluated for LEAFS in terms of detection limits and linear dynamic ranges. These atomizers include open atomization devices such as graphite cup and rod ETAs,[28,41,46,47,49,50,71,84,109,116,123–126,128,142,143] and enclosed atomization cells such as graphite tube ETAs.[39,40,52,53,73,74,77,122] For both cup and rod atomizers, the analyte atomic fluorescence has been observed above the atomizer surface in the right-angle mode. For the tube atomizers, both right-angle and front-surface detection, (Figure 3.6) modes have been employed to collect the fluorescence. In order to use right-angle detection for LEAFS, the tube atomizer must be modified by the addition of ports to allow passage of the laser radiation at a right angle to the fluorescence axis, but this leads to loss of atomic vapor. Conventional tube atomizers work well for LEAFS when used with front-surface illumination.[39,40,43,44,144]

An electrothermal atomizer has been used in two-step laser-excited atomic fluorescence [Figure 3.2g] for the determination of mercury.[145] The fluorescence collection was made at 90° by use of a pierced mirror, an achromatic lens, and a long-pass optical filter. A two-step excitation process at 253.7 and 435.8 nm was used, and direct fluorescence was observed at 546.1 nm to give an LOD of 90 fg (9 ppt with 10 μL injection). The linear dynamic range was five orders of magnitude. Indirect fluorescence collected with a less sensitive line at 407.8 nm allowed concentrations of above 1 ppm to be measured, which extended the LDR of the technique to at least seven orders of magnitude.

Double-resonance excitation fluorescence (Figure 3.2f) in a graphite tube electrothermal atomizer has been studied by Vera et al.[146] Two dye lasers pumped by a Nd:YAG laser were used to excite atoms in the atomizer simul-

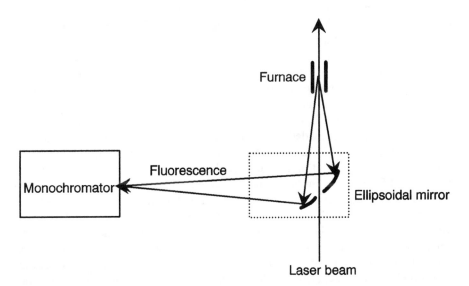

**Figure 3.6**  Front-surface illumination and detection of fluorescence from a graphite furnace.

taneously. The LODs for gallium, indium, and thallium were 2, 1, and 220 fg, respectively.

Zybin et al.[83] presented work on simultaneous multi-element analysis in a commercial graphite furnace atomizer with several CW semiconductor diode lasers. The fluorescence was detected by a simple photodiode without the use of dispersive units such as interference filters or polychromator, because the beams of both lasers were alternately chopped by a low-frequency chopper (about 10 Hz). LODs of 10 fg and 20 fg were obtained for lithium and rubidium, respectively. Further improvements could be expected from the application of diode lasers with higher CW power, optimization of the detector, and improved signal processing.

A side-heated graphite tube furnace has been used for the detection of vanadium at picogram levels by LEAFS.[147] The determination involved double-resonance excitation in the visible light region, followed by detection of fluorescence in the ultraviolet region by a solar-blind photomultiplier tube, which eliminated the influence of scattered laser light in the system. An LOD of 10 pg was achieved, which was a great improvement compared to the LOD of 1.7 ng reported in the literature[54] in which a graphite cup was used for ETA-LEAFS. The limiting factor for the detection of vanadium was contamination from the graphite furnace.[147]

Bolshov et al.[50,148,149] demonstrated that vacuum electrothermal atomizer–LEAFS (VETA-LEAFS) was successful for the elimination of matrix interferences and increased atomization efficiency without the use of the stabilized temperature platform furnace (STPF) technique. Lonardo et al.[150] attempted to improve VETA-LEAFS through the use of STPF technology, for the determination of phosphorus and tellurium in nickel alloys, cobalt in glass, and platinum in aqueous solution. The combination of VETA-LEAFS and STPF was not found to be beneficial. A reduction in working pressure resulted in increased analyte diffusion and, ultimately, a decreased sensitivity, while it did not greatly improve the atomization efficiency of the analyte. However, working pressures lower than one atmosphere may be useful in instances where a shift in the linear dynamic range or better temporal resolution are required.

A comparison of the best reported ETA-LEAFS detection limits for 33 elements with those provided by ETA-AAS[151] is shown in Table 3.1.[39–41,43,45,52,54,74,83,128,142,145,152–156] Elements that have poor detection limits in ETA-AAS because of furnace atomization problems, also have relatively poor ETA-LEAFS detection limits. For many elements, improvements of two to four orders of magnitude have been obtained because of the improved signal-to-noise ratio produced by the fluorescence technique with its low background and the high irradiance of the laser. All of the elements for which the ETA-LEAFS detection limits are relatively poor require very high atomization temperatures (>2500°C). The blackbody emission from the atomizer probably controls the detection limit for these elements. Recent improvements in optics, which will be discussed later, will probably improve these LODs. In some instances, the

**Table 3.1**  Comparison of LODs of ETA-LEAFS with Those of ETA-AAS

| Element | ETA-LEAFS LODs (fg) | ETA-AAS LODs (pg) | References |
|---------|---------------------|-------------------|------------|
| Ag | 8 | 0.5 | 52 |
| Al | 100 | 4 | 45 |
| Au | 3 | 10 | 156 |
| Ba | $4 \times 10^4$ | 10 | 128 |
| Bi | 2.5 | 10 | 155 |
| Cd | 18 | 0.3 | 39 |
| Co | 4 | 2 | 153 |
| Cs | 1000 | 5 | 142 |
| Cu | 150 | 1 | 154 |
| Eu | 300 | 10 | 154 |
| Fe | 70 | 2 | 45 |
| Ga | 1 | 40 | 152 |
| Hg | 90 | — | 145 |
| In | 2 | 9 | 152 |
| Ir | 2000 | 300 | 71 |
| Li | 10 | 2 | 83 |
| Mn | 100 | 1 | 45 |
| Mo | 100 | 4 | 40 |
| Na | 60 | 1 | 128 |
| Ni | 10 | 10 | 168 |
| P | 8 | 3000 | 43 |
| Pb | 0.2 | 5 | 45 |
| Pd | 700 | 25 | 54 |
| Pt | 100 | 50 | 41 |
| Rb | 20 | 5 | 83 |
| Rh | 2 | — | 54 |
| Sb | 10 | 15 | 166 |
| Sn | 30 | 20 | 45 |
| Te | 20 | 10 | 166 |
| Tl | 0.3 | 10 | 45 |
| V | 10 pg | 20 | 147 |
| Yb | 220 | — | 152 |

LEAFS detection limits have been improved by changing the furnace purge gas from argon to a mixture of hydrogen in argon to provide a reducing environment.[41]

The linear calibration range of ETA-LEAFS is four to seven orders of magnitude for most elements. The linear calibration ranges are usually better than those observed in AES, and superior to those obtained with atomic absorption, for which the linear range is limited to two to three orders of magnitude. Curvature of ETA-LEAFS calibration graphs at high concentrations has been attributed to pre- and postfilter effects for right-angle detection of fluorescence, as discussed earlier.[35] Filter effects seem to be largely nonexistent when front-surface illumination and detection are used.[45]

### 3.3.2.4  Glow Discharge

Although the glow discharge (GD) has been used frequently in mass spectrometry, atomic emission, and atomic absorption, little research has been done concerning its use in LEAFS.[157] A solid sample volatilized in a GD is present mainly as a vapor cloud of free atoms. The GD offers several advantages when used as an atom source.[158] First, an almost ideal environment for atomic fluorescence measurements can be achieved because of the low pressure and inert gas atmosphere. The low pressure reduces absorption line broadening, and the inert gas minimizes quenching of the fluorescence. The atomization of solid samples minimizes the sample preparation steps and reduces the chance of contamination from dissolution of the sample. An alternative method involves the drying of an aqueous sample on to an electrode prior to analysis in the GD.

For atomic fluorescence, the GD can be operated in a direct current mode or a pulsed mode. In the pulse mode of operation, the fluorescence is observed while the discharge is off in order to minimize background emission.[120] The noise was found to decrease by a factor of 10–100 when the discharge was off. Detection limits were reported for indium and lead, with direct current and pulsed operation of the discharge, respectively. Indium was determined with a detection limit of 10 ng (1 µg/mL) in aqueous solution and 8 ng in a solid,[121] while the detection limit for lead was 20 pg (4 ng/mL) in aqueous solution and 0.1 µg/g in a solid.[120]

Womack et al.[92] studied the application of a pulsed GD as an atomizer for atomic fluorescence excited with a CVL-pumped dye laser. Lead and indium displayed linear calibration curves with LODs of 15 fg and 2 fg, respectively. Work on the use of low-pressure discharges as atom sources for LEAFS has been reported.[159] This involved using a hot-hollow cathode source for the atomization of cobalt. The background signal was reduced by making fluorescence measurements in the 300 ms after the discharge had been switched off.

A planar cathode GD has been employed to determine lead.[160] Temporal and spatial profiles of the atom distribution showed that rapid sputtering and diffusion distributed lead throughout the sputtering chamber, with maximum fluorescence occurring 100 ms after the start of the sputtering discharge. The best signal-to-noise ratio was obtained from measurements just below the anode. The detection limit was 2 pg with sample volumes between 0.1 and 1.0 µL.

### 3.3.3  Optics

Generally, there are five main requirements for the development of the optical arrangement for a fluorescence detection system:[161] (1) the optical arrangement must have maximum collection efficiency over a wide wavelength range (195–800 nm); (2) it must discriminate between background and analytical signals without significant sacrifice in the latter; (3) there must be minimal losses of analytical signal along the optical path due to the reflection from different surfaces and apertures; (4) the optics should not be difficult to align; and (5) the components should be relatively inexpensive and commercially available.

For early experiments with flame LEAFS, several different optical arrangements were employed. The most simple one consisted of a biconvex lens[88,162–164] or a plane mirror.[4] Alternative arrangements consisted of systems of mirrors and had much higher collection efficiency and less aberration.[160,165] However, their disadvantages include difficulties with optical alignment and the absence of inexpensive and standardized optical components. A torroidal mirror integrated with biconvex lenses has been used for the enhancement of fluorescence signals by multiple light passes through the atomizer.[37,56,115] Recent improvements in collection efficiency for ETA-LEAFS have almost all involved front-surface illumination.[39,166]

One recent attempt to develop a quantitative model for the collection of fluorescence radiation in ETA-LEAFS was performed by Farnsworth et al.[167] For front-surface illumination, the authors compared optical collection efficiencies for biconvex, plano-convex, and achromatic lenses and concluded that the most aberration-free arrangement consisted of one flat mirror and a pair of achromatic lenses, while a pair of plano-convex lenses provided a useful compromise in cost and efficiency.

Yuzefovsky et al.[161] took advantage of the low optical aberrations for image transfer that occur with the use of an ellipsoidal mirror in a dispersive detection system. The lower optical aberration allowed more efficient spatial discrimination between the fluorescence and blackbody radiation, and led to better LODs. The best detection limit for cobalt with the new optical arrangement was 20 fg, which was a factor of five better than that obtained with conventional optical arrangements with otherwise the same instrumentation. The signal-to-background ratio and the fluorescence collection efficiency were studied as a function of position of the optical components. For both cobalt and phosphorus, the signal-to-background ratio with the ellipsoidal mirror was stable within 10–20% during ± 8 mm shifts in the position of the detection system from the focal plane of the optics.

### 3.3.4 Detection Systems and Background Correction

#### 3.3.4.1 Detection Systems

Detection systems usually consist of a monochromator or narrow-band optical filters tuned to an appropriate detection wavelength, a photomultiplier tube (PMT), and a boxcar averager.

For nondispersive detection of fluorescence, the effect of the bandpass of the filter on the signal-to-noise ratio (SNR) is significant.[45] A decrease in bandpass should result in a reduction of noise and an improvement of detection limit. In the determination of thallium, Wei et al.[45] showed that the detection limit of thallium was improved from 42 fg to 3 fg, and 0.3 fg for a broadband filter, a 10-nm bandpass filter, and a 1-nm bandpass filter, respectively. While there is no improvement in the SNR by varying the slit width for dispersive detection of fluorescence for either flicker noise–dominated or shot noise–dominated situations, a square root improvement in the SNR can be achieved by use of a larger slit width for the nondispersive, shot noise–dominated situation.

PMTs have a long history and are used primarily for the detection of atomic fluorescence because of their high sensitivity, low price, and ease of use.[168] On the other hand, there exist some problems in the use of a PMT as the detector. In LEAFS there is a very wide range of atomic fluorescence signal size. The PMT operating voltage is usually adjusted to reduce the current flowing in the dynode chain, to keep the PMT within its own linear operating range. However, when a high-power pulsed dye laser is used for the excitation, there can be a problem because the signals during each pulse are very large. The situation is made worse in any atom cell by broadband background which alone may cause the PMT to operate close to its maximum average current output, thus being unable to deliver high pulsed currents. This behavior is known as pulse-mode saturation and occurs when pulsed light signals are detected in the presence of an intense, continuous background.[169] Pulse-mode saturation in LEAFS occurs not only at high concentrations, but also at concentrations close to the detection limit. Although very wide monochromator slit widths can be employed to improve SNR when very low concentrations are encountered, more background light will be detected by the PMT at the same time. The first approach to these problems is to use charge storage capacitors in the dynode chain of the PMT.[85] Second, it is possible to make the charge storage capacitors more effective by use of PMT gating which switches the PMT off over 99% of the time and prevents it from seeing the background when it is unnecessary to do so. Seltzer et al.[94] employed PMT grating to improve SNRs in LEAFS with the use of large monochromator slits. Linear dynamic range was not improved by the use of gated PMT operation but was maintained at wider slit widths.

The charge-coupled device (CCD) has been used for the detection of fluorescence in LEAFS.[168,170] Marunkov et al.[168] and Enger et al.[170] explored this technique and employed it for the determination of nickel, aluminum, and lead. In their experiments, the laser radiation used for the excitation was produced by an excimer-pumped dye laser system. The fluorescence radiation was collected by use of the front-surface illumination mode. The spectrometer was equipped with a UV-enhanced intensified CCD (ICCD) as a detector, which was placed at the output focal plane of the spectrometer. The ICCD detector was operated at $-35°C$ to reduce thermally produced charges. The intensifier was gated for 110 ns directly following the laser pulse to collect the fluorescence radiation.

Compared with conventional PMT detection, the advantages of the ICCD approach are as follows: (1) several wavelengths can be detected simultaneously, which allows for rapid investigation of the atomic fluorescence spectra to find the most sensitive excitation and detection wavelength combination; this is very important for the analysis of elements with complex atomic structures, such as vanadium, chromium, manganese, iron, cobalt, nickel, and so forth; (2) improved absolute sensitivity can be achieved by simultaneous detection of fluorescence at several wavelengths; (3) improved spectral selectivity can be obtained; for example, the background signals resulting from scattering light, stray light and blackbody radiation can be monitored and corrected; (4) the possibility of time-resolved studies of fluorescence spectra exists; and (5) two-dimensional imaging of height distributions of atomization and diffusion processes in the atom cell can be obtained. In contrast, the disadvantages of the ICCD over the PMT include the somewhat lowered sensitivity that results from a lower quantum efficiency of the intensifier's photocathode, high cost, and limited readout rate. More research needs to be done to explore the utility of array detectors for laser-based atomic fluorescence.

### 3.3.4.2 Background Correction

Background signals carry the noise discussed earlier and may arise from scattered and stray radiation, atom cell emission, molecular fluorescence, and radio frequency interference. In addition, background may result from spectral overlap, which does not often occur in LEAFS. Among these sources, scattered and stray radiation and atom cell emission are the main sources of background for LEAFS.

Several background correction techniques have been reported for flame LEAFS and ETA-LEAFS, including the use of a nonanalyte line, nonresonance fluorescence, wavelength modulation, Zeeman background correction, intermodulated fluorescence and harmonic saturation spectroscopy, two-channel boxcar detection, time-resolved fluorescence, multichannel background correction, variation of the excitation spectral profile, and so on. In this section, only wavelength modulation and Zeeman background correction techniques are discussed because they are of the most practical significance for ETA-LEAFS. For other background correction methods, readers are referred to review articles.[25,27]

In the wavelength modulation (WM) method, the laser output is alternated between the analytical wavelength for the measurement of atomic fluorescence and background, and some wavelength slightly higher or lower than the analytical wavelength for the measurement of the background alone. Su et al.[44] and Irwin et al.[76] have reviewed the concept of laser wavelength modulation. Su et al.[44] employed a piezoelectric pusher to drive the wavelength tuning mirror in a laboratory-constructed grazing incidence dye laser. The laser pulses were synchronized with the piezoelectric pusher movement so that alternate laser pulses measured the atomic fluorescence signal at the analytical atomic spectral line (on line) and the background signal at a wavelength displaced to one side of the atomic line (off line). The background-corrected signal was obtained by subtracting the off-line "background" from the on-line "signal plus background." Compared with other LEAFS background-correction techniques, wavelength modulation offers several advantages. It is easy to obtain wavelength-modulated laser light with a piezoelectric pusher, and it provides high temporal resolution and high spectral resolution. Wavelength modulation corrects all the major background problems such as scatter of laser radiation off concomitant species, stray light, flame emission in flame LEAFS, and blackbody emission in furnace LEAFS, and it should also correct broadband molecular fluorescence background. The only disadvantage of this technique is that it does not correct background exactly at the same wavelength as the analytical line, which may result in some error when structured background or spectral interferences exist near or within the analytical line. Fortunately, both of these seldom appear in LEAFS because of its inherent high spectral selectivity.

Figure 3.7 shows Zeeman energy levels for an alternating current magnetic field parallel to the excitation beam, assuming that there is a normal Zeeman splitting pattern. When the magnetic field is off, the analyte atomic fluorescence signal plus background is measured; when the magnetic field is on, the analyte atomic energy levels are split into sigma components displaced away from the analytical wavelength, and only background is measured. The "field off" and the "field on" signals can be separately processed by a boxcar averager, and their difference can be obtained by the boxcar to provide the background-corrected atomic fluorescence measurement.

The Zeeman background correction technique is probably the most effective method for LEAFS.[40,76,77,171] Background that results from scattered or stray radiation, blackbody emission, molecular fluorescence, and radio frequency noise can be corrected by Zeeman background-corrected ETA-LEAFS (ZETA-LEAFS). Compared with other background correction techniques, the advantages of the Zeeman technique are that the background correction measurement

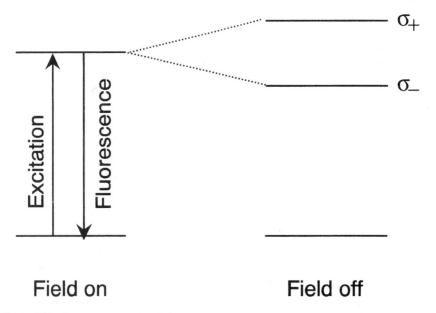

**Figure 3.7**   Zeeman energy levels for an alternating current magnetic field parallel to the excitation beam.

is made at the analytical wavelength and that only one radiation source is required. ZETA-LEAFS has a long linear dynamic range (five to seven orders of magnitude) and high sensitivity, and the LODs for real samples are not significantly degraded relative to pure aqueous solution. However, molecular spectra that contain fine detail may interfere with Zeeman background correction.

### 3.3.5　Isotope Determinations

#### 3.3.5.1　ICP-LEAFS Approach

Recently, Vera et al.[139] reported the extension of ICP-AFS to the determination of concentrations of uranium isotopes and the first experimental demonstration of isotopically-resolved ICP-LEAFS. This finding is of general analytical importance since it suggests that the powerful isotopic dilution technique may be useful with ICP-LEAFS. The uranium spectrum is characterized by large isotope splitting, which allows the measurement to be made with a pulsed laser that has a relatively wide spectral linewidth. Further development of this technique awaits improvements in laser technology to produce pulsed lasers with a narrow spectral linewidth of about 1 MHz or better and high pulse energy. CW lasers can provide narrow single-mode spectral linewidth but have not, so far, been able to provide the wavelength versatility of pulsed lasers. The LOD for uranium via ICP-LEAFS using the 286.57/288.96-nm excitation/fluorescence line pair was estimated to be about 2 µg/mL. The nonresonance fluorescence determination of uranium dissolved in a very complex matrix was shown to be independent of spectral line interference from the matrix elements.

The Doppler broadening present in the ICP prevents the determination of isotopes by ICP-LEAFS when their spectral lines have isotope shifts that are smaller than the broadened linewidths. Only the lightest and heaviest elements have isotope shifts sufficiently large to overcome the ICP-source broadening. Methods such as saturation spectroscopy and two-photon spectroscopy can, in principle, produce spectra that are not Doppler broadened, and may extend ICP-LEAFS isotopic measurements to a greater number of elements.

The selectivity of ICP-LEAFS depends on the spectral linewidth of the laser excitation source and can be expected to be improved as reliable, pulsed, single-mode lasers are developed. Presently, the LODs of isotopically selective LEAFS are poor compared with those of high-resolution ICP-AES, nonisotopically resolved ICP-LEAFS, and ICP-MS.

#### 3.3.5.2　Doppler-Free Two-Photon Spectroscopy

Hyperfine isotope spectra are usually buried by Doppler broadening. To obtain isotope-discriminated spectral peaks, Doppler broadening must be eliminated. Doppler-free two-photon spectroscopy, which occurs when two narrow bandwidth and high-power laser beams illuminate the same atom simultaneously, could be used to obtain isotope-discriminated spectra.[79,172]

In an experiment of Doppler-free two-photon spectroscopy, the two overlapping laser beams could be arranged in co- or counterpropagating directions through an atom cell. In either case, the narrow bandwidth first laser excites only a small group of atoms with velocity $V_x$, which satisfies the Doppler shift condition:

$$\nu_{laser1} = \nu_{atom}(1 \pm V_x/c) \tag{3.5}$$

Here, $\nu_{laser1}$ is the first laser frequency, $\nu_{atom}$ is the atomic resonance frequency, $V_x$ is the linear velocity of the atoms, and $c$ is the velocity of light. Simultaneously, the tunable, high-power second laser interacts with this group of excited atoms with the same linear velocity, $V_x$, and pumps the atoms to an even higher energy level.

If the two laser beams have the same frequency and are arranged in counterpropagating directions, equal and opposite Doppler shifts will result in a zero net shift. Thus, the resultant sum frequency is constant and independent of the particle's velocity. Therefore, it is possible to obtain Doppler-free isotope-discriminated spectra.

If the two laser beams propagate from the same direction or with different frequencies, so-called two-color Doppler-free two-photon spectroscopy, a Doppler shift will exist, but Doppler broadening will be eliminated because the first laser picks up only those atoms with the same linear velocity, $V_x$. In these cases, Doppler-free isotope-discriminated spectra will be obtained with a Doppler shift, as seen from equation (3.5).

In Doppler-free two-photon spectroscopy, the intensity of the lasers and density of atoms are two main factors that determine the likelihood of the simultaneous absorption. However, the intermediate quantum state between the ground and the excited states could be either an existent state or an imaginary one. If the real intermediate quantum state is absent, the population probability of the excited states is so low that it is difficult to reach saturation of the two-photon transition. If an intermediate state exists, the two-photon excitation probability increases by orders of magnitude, and saturation might be taken into account.

Niemax et al.[173–175] have investigated the potential of the determination of isotopes and isotope ratios by resonant Doppler-free two-photon laser-enhanced ionization spectroscopy. The basic experimental arrangement is shown in Figure 3.8. The overlapping beams of two single-mode CW lasers were directed simultaneously in counterpropagating directions through two cells that contained a noble gas. The element under investigation, calcium, was evaporated continuously in a reference cell, while the sample was atomized in the analyte cell. The potential of isotope peaks discrimination and isotope ratio measurements by this technique was demonstrated.

Doppler-free two-photon-excited fluorescence spectroscopy of OH in flame has been studied by Goldsmith and Rahn[176] (Figure 3.9). An argon laser was used to pump a CW ring dye laser system that provided 5-mJ, 22-nsec pulses at 620 nm, with a measured bandwidth of 40 MHz. An 18-inch focal-length lens

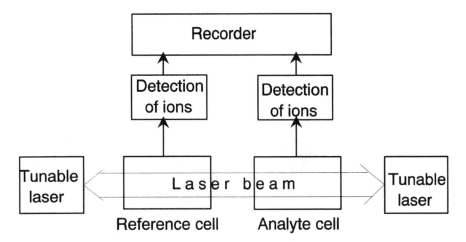

**Figure 3.8**   General experimental setup for resonant Doppler-free two-photon spectroscopy with two lasers. Reproduced with permission from reference 175.

was used to focus the laser beam into the low-pressure flame chamber, and the laser beam was reflected back by a plane mirror. A Faraday-type optical isolator was used to isolate the laser system from the retroreflected laser pulses. Following two-photon excitation by OH in the flame, the 310-nm fluorescence emission was collected by a PMT. The gain in spectral resolution obtained by use of the Doppler-free excitation is shown in Figure 3.10. The lower curve was recorded in a 20-torr flame with the retroreflected beam present, leading to Doppler-free excitation.

## 3.4  Applications

In real sample analysis by LEAFS, both solid sampling and aqueous solution sampling have been used. In addition, molecular fluorescence of small molecules can be excited and measured by use of the same LEAFS instrumentation, with a view to the determination of nonmetals. In this section, solid sampling, real sample analysis by LEAFS, and laser-excited molecular fluorescence spectrometry (LEMOFS) are discussed.

### 3.4.1  Solid Sampling

Solid sampling can be done with direct insertion devices, electrothermal vaporizers, chemical vapor generation, and laser ablation. Solid sampling offers several possible advantages, including the reduction of sample pretreatment processes, minimization of sample pretreatment time, and avoidance of the potential contamination by reagent impurities. Furthermore, the chance of in-

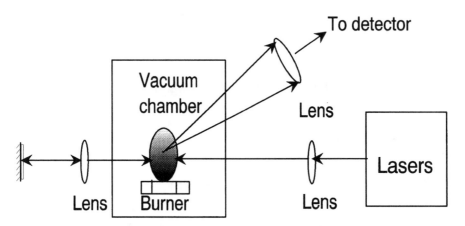

**Figure 3.9**   Experimental setup for Doppler-free two-photon spectroscopy.

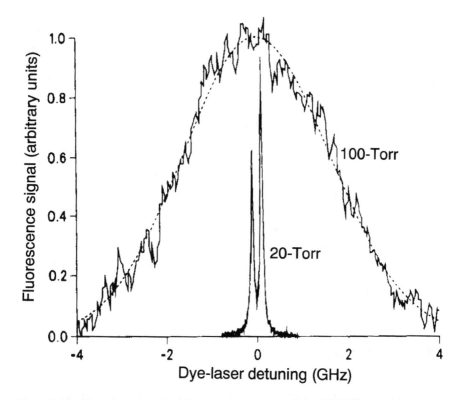

**Figure 3.10**   Two-photon-excited fluorescence spectra of the OH $S_1(5)$ transition: upper curves, Doppler-broadened spectrum in a 100-torr flame (*solid curve*) and least-squares fit to a Gaussian profile (*dashed curve*); lower curve, Doppler-free spectrum in a 20-torr flame. Reproduced with permission from reference 176.

terference from acid anions and loss of volatile elements during digestion procedures can be greatly reduced.

LEAFS with sample atomization by ion sputtering (SAIS) in a planar magnetron has been shown to be a potentially powerful technique for the direct analysis of impurity traces in pure solid materials.[177] The potential of LEAFS-SAIS was examined through the analysis of silicon contents in pure indium and gallium. Both theoretical and experimental calibration curves were constructed. The LODs obtained for silicon in indium and gallium were 0.4 ng/g and 1 ng/g, respectively.

The direct analysis of solid nickel-based alloys has been investigated by use of ETA-LEAFS with front-surface illumination and a Perkin-Elmer cup solid sampling accessory.[144] The results showed five to seven orders of magnitude LDR and one to four orders of magnitude better LOD than ETA-AAS. The analyte backgrounds for solid sampling by ETA-LEAFS were significantly lower or nonexistent compared to those for solid sampling by ETA-AAS. Trace metals thallium and lead in high-temperature nickel-based alloy standard reference materials were accurately determined by aqueous calibration.

Liang et al.[166] determined the metalloids tellurium and antimony in nickel-based alloys by ETA-LEAFS with direct solid sampling. The detection limits were three orders of magnitude better than those obtained by GF-AAS. The LDRs were found to be six and seven orders of magnitude for antimony and tellurium, respectively, and aqueous calibration could be used for the direct solid sample analysis.

Lonardo et al.[178] determined phosphorus in polymers that consisted mainly of poly(ethylene terephthalate) by use of ETA-LEAFS with direct solid sampling. The results were in good agreement with those of ETA-AAS and ICP-AES. Of the three techniques, only ETA-LEAFS could measure the phosphorus content of all the samples over a wide range of concentrations between 2 and 3000 µg/g.

The earliest work on laser-ablation fluorescence spectrometry appears to have been done by Measures et al.[179] to measure the radiative lifetimes of three resonance transitions of chromium. They used a Q-switched ruby laser and a nitrogen-pumped dye laser for sample ablation and excitation, respectively. A trace element analyzer based on laser ablation and selectively excited radiation (known as TABLASER) was developed.[180,181] The sample was contained in a low-pressure sample chamber that allowed the laser-produced plasma to expand away from the sample. Chemical interferences in TABLASER were studied during analysis of standard reference materials such as milk powder, flour, and steel samples.[181,182] It was shown that the advantages of TABLASER included selective microsample capability, no sample preparation, real-time analysis, high sensitivity, freedom from matrix effects, high selectivity, wide LDR, depth-profiling capability, and the possibility of simultaneous multi-element measurement.

The analytical properties of the plume from laser-ablated steel, copper, and aluminum samples have been studied by Niemax and coworkers.[177,183,184] A pulsed Nd:YAG laser (5 J per pulse) was used for sample ablation, while an excimer-pumped dye laser was used as the excitation source. Optimum analytical fluorescence signals were obtained with 14 kPa of argon atmosphere. Readers, for more information, are referred to reviews on the interaction of laser radiation with solid materials and laser microanalysis.[185,186]

### 3.4.2 Real Sample Analysis

LEAFS has been applied to analyze real samples such as alloys, water, polymers, and biological samples. Elements determined include cadmium, cobalt, copper, iron, manganese, sodium, nickel, lead, thallium, tellurium, tin, strontium, and so forth. Table 3.2 summarizes the analytical results for real samples analyzed by LEAFS. Among those atom cells discussed previously, the graphite furnace has been the most popular atomizer for LEAFS in real sample analysis. Some work has been done with flames, but relatively few real samples have been analyzed with ICP or GD atomizers. Overall, the numbers of both samples and elements analyzed by LEAFS are limited, and more work needs to be done. Real sample analyses have been discussed in some review articles.[13,25,187]

Flame LEAFS has been used as a high-sensitivity detector for organomanganese and organotin compounds following separation by HPLC.[42] The HPLC-flame LEAFS instrumentation was applied to an investigation of the manganese species responsible for (methylcyclopentadienyl)manganese tricarbonyl (MMT) toxicity in rats. The detection limits for various organomanganese species by HPLC-flame LEAFS ranged from 8 to 22 pg of manganese. Recovery of these compounds from rat urine varied from 80 to 100%, with an RSD of between 4 and 8%.

Sodium in NIST standard reference materials (SRM) was determined with wavelength-modulated flame LEAFS.[44] Two-directional wavelength modulation provided accurate correction for both nonstructured background and sloping background at the analytical line.

Manganese in zinc chloride matrix and in mouse brain tissue has been determined by LEAFS in an electrothermal atomizer with Zeeman background correction.[78] This work was done in a furnace with additional holes for right-angle detection of the fluorescence. As a result, furnace conditions did not adhere to STPF technology. Accordingly, the method of standard addition was used for the determination of manganese in mouse brains. ZETA-LEAFS was shown to correct for background for the resonance fluorescence determination of manganese in aqueous solution, zinc chloride matrix, and mouse brain tissue.

A CVL-pumped dye laser has been used in the direct determination of lead in Great Lakes water and certified reference materials.[188] A working detection limit of 0.4 ng/L with 25 µL injection volume (10 fg absolute) was obtained.

Thallium in NIST (National Institute of Science and Technology, Gaithersberg, MD) biological samples has been accurately determined by capacitive discharge graphite furnace LEAFS.[189] The LOD found was one to two orders of magnitude superior to that of ETA-AAS. The interference effects of calcium, sodium chloride, and potassium chloride on the thallium signal were shown to be similar to both conventional and capacitive discharge furnaces LEAFS. The

**Table 3.2**  Real Sample Analyses by LEAFS

| Elements | Samples* | LEAFS results | Certified results | References |
|---|---|---|---|---|
| Co | Sea water (CASS-2) | 22 ± 2 | 25 ± 6 ng/L | 141 |
| Co | Vegetation samples | | | 50,148 |
| | (SBMP-01) | 45 ± 10 | 60 ± 20 µg/g | |
| | (SBMK-01) | 80 ± 15 | 100 ± 30 µg/g | |
| | (SBMT-01) | 65 ± 10 | 60 ± 20 µg/g | |
| Cu | Wheat flour (SRM 1567) | 2.5 | 2.0 ± 0.3 µg/g | 41 |
| | Spinach (SRM 1570) | 17 | 12 ± 2 µg/g | |
| | Steel (SRM 364) | 0.29 | 0.24% | |
| Fe | Water (SRM-1643) | 78 | 75 ± 1 ng/g | 56 |
| | Unalloyed copper (SRM-394) | 145 ± 6 | 147 ± 8 µg/g | |
| | Fly ash (SRM 1633) | 6.2 ± 0.2 | 6.2 ± 0.3 µg/g | |
| Fe | Pine needles (SRM 1575) | 198 ± 8 | 200 ± 10 µg/g | 113 |
| Mn | Zinc chloride | 5.0 ± 1.5 ng/mL | NA | 78 |
| | Mouse brain | 1.9 ± 0.4 ng/mL | NA | |
| Mn | Wheat flour (SRM 1576) | 7.3 | 8.5 ± 0.5 µg/g | 41 |
| | Spinach (SRM 1570) | 128 | 165 ± 6 µg/g | |
| | Steel (SRM 364) | 0.23 | 0.23% | |
| Mn | Citrus leaves (SRM 1572) | 21.5 ± 0.5 | 23 ± 0.2 µg/g | 191 |
| | Bovine liver (SRM 1577a) | 10.5 ± 0.3 | 9.90 ± 0.8 µg/g | |
| | Milk powder (SRM 1549) | 0.28 ± 0.01 | 0.26 ± 0.06 µg/g | |
| Organo manganese | Rat urine | 80–100% recovery | NA | 42 |
| Na | Bovine (SRM 1577a) | 0.251 ± 0.014 | 0.243 ± 0.013% | 44 |
| | Nonfat milk powder (SRM 1549) | 0.454 ± 0.047 | 0.497 ± 0.010% | |
| | Citrus leaves | 150 ± 14 | 160 ± 20 µg/g | |
| Ni | Water (SRM 1643) | 50 ± 2 | 49 ± 1 ng/g | 108 |
| | Unalloyed copper (SRM 396) | 4.1 ± 0.1 | 4.2 ± 0.1 µg/g | |
| | Fly ash (SRM 1633) | 99 ± 3 | 98 ± 3 µg/g | |

(continued)

**Table 3.2**   Real Sample Analyses by LEAFS (Continued)

| Elements | Samples | LEAFS results | Certified results | References |
|---|---|---|---|---|
| Pb | Water (SRM 1643c) | 36.4 ± 1.5 | 35.3 ± 0.9 ng/L | 188 |
|  | Water (SLRS-2, NRC) | 125.3 ± 10.8 | 129 ± 11 ng/L |  |
| Pb | Nickel-based alloy (SRM 897) | 12.6 ± 1.2 | 11.7 ± 0.8 µg/g | 144 |
|  | (SRM 898) | 2.2 ± 0.4 | 2.5 ± 0.6 µg/g |  |
|  | (SRM 899) | 3.7 ± 0.7 | 3.9 ± 0.1 µg/g |  |
| Pb | Estuarine sediment (SRM 1646) | 28.4 ± 3.2 | 28.2 ± 1.8 µg/g | 76 |
|  | Coal fly ash (SRM 1633a) | 74.4 ± 7.9 | 72.4 ± 0.4 µg/g |  |
|  | Citrus leaves (SRM 1572) | 13.2 ± 1.7 | 13.3 ± 2.4 µg/g |  |
|  | Pine needles (SRM 1575) | 10.3 ± 0.8 | 10.8 ± 0.5 µg/g |  |
| Pb | Citrus leaves (SRM 1572) | 14.4 ± 0.6 | 13.3 ± 2.4 µg/g | 191 |
| Pb | Water (SRM 1643) | 20 ± 2 | 20 ± 1 µg/g | 113 |
| Pb | Unalloyed copper (SRM 394) | 26 ± 2 | 26.5 ± 0.2 µg/g | 113 |
| Sn | Unalloyed copper (SRM 394) | 66 ± 3 | 65 ± 5 µg/g | 108 |
| Sr | Pine needles (SRM 1575) | 4.7 ± 0.2 | 4.8 ± 0.2 µg/g | 113 |
| Te | Nickel alloy (SRM 897) | 0.97 ± 0.15 | 1.05 ± 0.07 µg/g | 166 |
|  | (SRM 898) | 0.52 ± 0.05 | 0.54 ± 0.02 µg/g |  |
|  | (SRM 899) | 5.4 ± 0.7 | 5.9 ± 0.6 µg/g |  |
| Tl | Bovine liver (SRM 1577a) | 2.9 ± 0.2 | 3.1 ± 0.1 ng/g | 53 |
| Tl | Nickel-based alloy (SRM 897) | 0.56 ± 0.07 | 0.51 ± 0.03 µg/g | 144 |
|  | (SRM 898) | 2.5 ± 0.5 | 2.75 ± 0.02 µg/g |  |
|  | (SRM 899) | 0.24 ± 0.03 | 0.252 ± 0.003 µg/g |  |
| Tl | Bovine liver (SRM 1577a) | 0.0027 ± 0.0004 | 0.003 µg/g | 189, 191 |
|  | Pine needles (SRM 1575) | 0.048 ± 0.004 | 0.05 µg/g |  |
|  | Tomato leaves (SRM 1573) | 0.05 ± 0.01 | 0.05 µg/g |  |

*SRM is a National Institute for Standards and Technology Standard Reference Material.
CASS-2 is a National Research Council, NRC, Canada, Analytical Chemistry Standard.
SBMP is a Russian Standard Reference Material.

naturally occurring level of thallium in mouse brains has also been determined by LEAFS with STPF technology.[53]

Trace amounts of lead and cadmium in Antarctic and Greenland ancient ice and recent snow have been determined directly by LEAFS.[140] The concentrations determined varied in the range of 0.1–3 pg/ml for cadmium and 0.3–30 pg/ml for lead. These results agreed with those determined by isotope dilution mass spectrometry and ETA-AAS, which required time-consuming pretreatment and preconcentration process. Lead and cobalt in standard reference materials have been determined by Zeeman background correction LEAFS.[76] Lead has been accurately determined in slurried samples of estuarine sediment, coal fly ash, and citrus leaves by simple aqueous calibration, with precision in the 6–14% range. Cobalt has been determined in coal fly ash and estuarine sediment by use of dissolved sampling and palladium matrix modifier, with sample precisions of 5–7%.

Trace amounts of vanadium in blood and serum have been detected by LEAFS with a side-heated graphite furnace.[147] It was found that the limiting factor for the determination of vanadium was contamination from the graphite furnaces. Ultratrace amounts of cobalt in sea water have been determined by ETA-LEAFS with flow-injection semi-on-line preconcentration.[141] The preconcentration allowed separation of most of the matrix elements.

Antimony in environmental and biological samples at trace levels has been determined by ETA-LEAFS with an ICCD detector.[190] The ICCD detector permitted the simultaneous multichannel detection of large fluorescence regions, and control and correction for various types of background. Both aqueous and solid samples were investigated with respect to their antimony content. There have been some other articles on real sample analyses by LEAFS, with analytical results included in Table 3.2.[41,50,108,113,148,191]

### 3.4.3  Laser-Excited Molecular Fluorescence Spectrometry

Laser-excited molecular fluorescence spectrometry (LEMOFS) has been used to determine small molecules by means of LEAFS instrumentation. The work on LEMOFS as an analytical method for fluoride, chloride, and bromide has been done by Dittrich et al.[192] and Anwar et al.[193–195] Liang et al.[43] determined phosphorus by use of the phosphorus monoxide molecules in plant and biological reference materials by LEMOFS. Phosphorus monoxide was excited at 246.291 nm, and its fluorescence was detected at 324.5 nm, with a phosphorus detection limit of 80 pg. Su and Michel[196] determined chlorine in biological standard reference materials through indium monochloride fluorescence, which was excited at 267.2 nm and detected at 269 nm by ETA-LEMOFS. A chloride detection limit of 10 pg was achieved. The effects of metal ions and nonmetal ions were studied. In general, for LEMOFS matrix interferences were present in the analyses, but they could be removed by dilution of the sample. The high sensitivity of LEMOFS allowed this to be a practical approach. By use of simple aqueous calibration, the results obtained for the NIST reference samples were in good agreement with the noncertified reference values of chlorine. Butcher et al.[197] reported significant improvements in the determination of fluorine in urine and tap water by use of LEMOFS of magnesium monofluoride in a graphite tube furnace. Excess of magnesium was added to the samples to promote the formation of the magnesium monofluoride. The method was very sensitive with a detection limit of 0.3 pg, and allowed low levels of fluorine to be determined in the urine samples by means of simple aqueous calibration. The use of barium as a chemical modifier increased the size of the signal by a factor of 100, while the role of barium as a chemical modifier was investigated by Yuzefovsky and Michel,[198] based on a thermodynamics point of view.

## 3.5  Conclusions and Future Prospects

LEAFS is a useful instrumental method for ultratrace element analysis because of its exceptionally high sensitivity and selectivity. ETA-LEAFS is sensitive enough to be a single-atom detection (SAD) technique, which has been discussed by Alkemade,[10,19] Winefordner et al.,[97,199] and Hieftje.[200] However, SAD has been achieved only in quartz atom cells with controlled vapor pressure of species such as sodium. In this laboratory, we are currently able to detect several to several hundred atoms by ETA-LEAFS after accounting for loss of atoms in the system.[45] There is still room for further improvement in LODs by improvement in atomization efficiency and excitation and detection efficiency. Enormous improvement in LODs (subfemtogram) for several elements has been achieved since the first report of LEAFS.

Similar to GF-AAS,[201,202] ETA-LEAFS has the potential for absolute analysis or semiquantitative absolute analysis,[13,203] but this is more complicated than GF-AAS absolute analysis. It is difficult to prepare standard solutions at the femtogram level, even in a "clean" laboratory. In addition, there are no certified reference materials available at such low concentration levels. Therefore, it is very important to study absolute analysis by ETA-LEAFS. In a study of the possibility of performing absolute analysis by LEAFS in a graphite furnace atomizer, Omenetto[203] evaluated and discussed various theoretical and experimental parameters needed for conversion of a measured fluorescence signal into the corresponding number of atoms.

Research on many aspects of LEAFS has been reviewed recently.[204] For real sample analysis, more work needs to be done because the variety of both samples and elements analyzed is limited. The involatile elements, which usually have relatively low sensitivity in ETA-AAS, have not been explored rigorously in tube furnaces. It remains to be seen how sensitively these elements can be determined by LEAFS compared to AAS.

Most LEAFS work has been done with dye lasers. These are inconvenient to operate and the dye needs to be changed for almost every element. Hopefully, these difficulties can be alleviated by the use of solid lasers such as the OPO laser. Meanwhile, the use of an OPO laser with a multichannel CCD as the fluorescence detector may make sequential multi-element determination possible. The use of diode lasers in LEAFS, which has already shown promise for simultaneous detection of multiple elements, also has a promising future.

# References

1. Winefordner, J. D.; Vicker, T. J. *Anal. Chem.* **1964**, *36*, 161.

2. Winefordner, J. D.; Staab, R. A. *Anal. Chem.* **1964**, *36*, 165.

3. Denton, M. B.; Malmstadt, H. V. *Appl. Phys. Lett.* **1971**, *18*, 485.

4. Fraser, L. M.; Winefordner, J. D. *Anal. Chem.* **1971**, *43*, 1693.

5. Kuhl, J.; Marowsky, G. *Opt. Commun.* **1971**, *4*, 125.

6. Butcher, D. J.; Dougherty, J. P.; McCaffrey, J. T.; Preli, F. R. Jr.; Walton, A. P.; Michel, R. G. *Prog. Anal. Atom. Spectrosc.* **1987**, *10*, 395.

7. Omenetto, N.; Winefordner, J. D. *Prog. Anal. Atom. Spectrosc.* **1979**, *2*, 1.

8. Hill, S. I.; Dawson, J. B.; Price, W. J.; Riby, P.; Shuttler, I. L.; Tyson, J. F.; *J. Anal. Atom. Spectrom.* **1994**, *9*, 213R.

9. *Analytical Laser Spectroscopy;* Omenetto, N., Ed.; Wiley: New York, 1979.

10. Alkemade, C. Th. J. In *Analytical Applications of Lasers;* Piepmeier, E. H., Ed.; Wiley: New York, 1986.

11. Alkemade, C. Th. J. *Spectrochim. Acta* **1985**, *40B*, 1331.

12. Omenetto, N.; Human, H. G. C. *Spectrochim. Acta* **1984**, *39B*, 1333.

13. Sjöström, S. *Spectrochim. Acta Rev.* **1990**, *13*, 407.

14. Winefordner, J. D.; *J. Chem. Ed.* **1978**, *55*, 72.

15. *New Applications of Lasers to Chemistry;* Hieftjie, G. M., Ed.; ACS Symposium Series 85; American Chemical Society: Washington, DC, 1978.

16. Falk, H. *Prog. Anal. Atom. Spectrosc.* **1980**, *3*, 181.

17. *Lasers in Chemical Analysis;* Hieftje, G. M.; Travis, J. C.; Lytle, F. E. Eds.; Human Press: Clinton, NJ, 1981.

18. Alkemade, C. Th. J.; Hollander, Th.; Snelleman, W.; Zeegers, P. J. *Metals Vapours in Flames;* Pergamon: London, 1982.

19. Alkemade, C. Th. J. *Appl. Spectrosc.* **1981**, *35*, 1.

20. Blackburn, M. B.; Mermet, J.-M.; Boutilier, G. D.; Winefordner, J. D. *Appl. Opt.* **1979**, *18*, 1804.

21. Smith, B. W.; Rutledge, M. J.; Winefordner, J. D. *Appl. Spetrosc.* **1987**, *41*, 613.

22. Piepmeier, E. H. *Spectrochim. Acta* **1972**, *27B*, 431.

23. Piepmeier, E. H. In *Atomic Absorption Spectrometry with Laser Primary Sources;* Omenetto, N., ed.; Wiley: New York, 1979.

24. Piepmeier, E. H. *Analytical Applications of Lasers;* Wiley: New York, 1986.

25. Butcher, D. J.; Dougherty, J. P.; Preli, F. R.; Walton, A. P.; Wei, G.-T.; Irwin, R. L.; Michel, R. G. *J. Anal Atom. Spectrom.* **1988**, *3*, 1059.

26. Boutilier, G. D.; Bradshaw, J. D.; Weeks, S. J.; Winefordner, J. D. *Appl. Spectrosc.* **1977,** *31,* 307.

27. Omenetto, N.; Winefordner, J. D. *Prog. Anal. Atom. Spectrosc.* **1985,** *8,* 371.

28. Omenetto, N.; Benetti, P.; Hart, L. P.; Winefordner, J. D.; Alkemade, C. Th. J. *Spectrochim. Acta* **1973,** *28B,* 289.

29. Piepmeier, E. H. *Spectrochim. Acta* **1972,** *27B,* 445.

30. Hosch, J. W.; Piepmeier, E. H. *Appl. Spectrosc.* **1978,** *32,* 444.

31. Omenetto, N.; Bower, J.; van Dijk, C. A.; Winefordner, J. D. *J. Quant. Spectrosc. Radiat. Transfer* **1980,** *24,* 147.

32. Sychra, V.; Svoboda, V.; Rubeska, I. *Atomic Fluorescence Spectroscopy;* van Nostrand-Reinhold: New York, 1975.

33. Winefordner, J. D.; Schulman, S. G.; O'Haver, T. C. *Luminescence Spectroscopy in Analytical Chemistry;* Wiley: New York, 1972.

34. Hooymayers, H. P. *Spectrochim. Acta* **1968,** *23B,* 567.

35. Dougherty, J. P.; Preli, F. R. Jr.; Wei, G.-T.; Michel, R. G. *Appl. Spectrosc.* **1990,** *44,* 934.

36. Omenetto, N.; Hart, L. P.; Benetti, P.; Winefordner, J. D. *Spectrochim. Acta* **1973,** *28B,* 301.

37. Weeks, S. J.; Haraguchi, H.; Winefordner, J. D. *Anal. Chem.* **1978,** *50,* 360.

38. Omenetto, N.; Winefordner, J. D. *Appl. Spectrosc.* **1972,** *26,* 555.

39. Omenetto, N.; Cavalli, P.; Broglia, M.; Qu, P.; Rossi, G.; *J. Anal. Atom. Spectrom.* **1988,** *3,* 231.

40. Goforth, D.; Winefordner, J. D. *Talanta* **1987,** *34,* 290.

41. Goforth, D.; Winefordner, J. D. *Anal. Chem.* **1986,** *58,* 2598.

42. Walton, A. P.; Wei, G.-T.; Liang, Z.; Michel, R. G. *Anal. Chem.* **1991,** *63,* 232.

43. Liang, Z.; Lonardo, R. F.; Takahashi, J.; Michel, R. G. *J. Anal. Atom. Spectrom.* **1992,** *7,* 1019.

44. Su, E. G.; Irwin, R. L.; Liang, Z.; Michel, R. G. *Anal. Chem.* **1992,** *64,* 1710.

45. Wei, G.-T.; Dougherty, J. P.; Preli, F. R. Jr.; Michel, R. G. *J. Anal. Atom. Spectrom.* **1990,** *5,* 249.

46. Bolshov, M. A.; Zybin, A. V.; Koloshnikov, V. G.; Koshele, K. N.; *Spectrochim. Acta* **1977,** *32B,* 279.

47. Bolshov, M. A.; Zybin, A. V.; Koloshnikov, V. G. *Kvantovaya Elektronica* **1980,** *7,* 1808.

48. Epstein, M. S.; Winefordner, J. D. *Prog. Anal. Atom. Spectrosc.* **1984,** *7,* 67.

49. Bolshov, M. A.; Zybin, A. V.; Koloshnikov, V. G.; Asnetsov, M. V. *Spectrochim. Acta* **1981,** *36B,* 345.

50. Bolshov, M. A.; Zybin, A. V.; Koloshnikov, V. G.; Mayorov, I. A.; Smirenkina, I. I. *Spectrochim. Acta* **1986,** *41B,* 487.

51. *Analytical Methods for Atomic Absorption Spectrometry;* Perkin-Elmer: Norwalk, CT, 1982.

52. Dougherty, J. P.; Preli, F. R. Jr.; Michel, R. G. *J. Anal. Atom. Spectrom.* **1987,** *2,* 429.

53. Dougherty, J. P.; Costello, J.; Michel, R. G. *Anal. Chem.* **1988,** *60,* 336.

54. Falk, H.; Tilch, J. *J. Anal. Atom. Spectrom.* **1987,** *2,* 527.

55. Smith, B. W.; Blackburn, J. D.; Winefordner, J. D. *Can. J. Spectrosc.* **1977**, *22*, 57.

56. Epstein, M. S.; Bayer, S.; Bradshaw, J.; Voigtman, E.; Winefordner, J. D. *Spectrochim. Acta* **1980**, *35B*, 233.

57. Omenetto, N.; Human, H. G. C.; Cavalli, P.; Rossi, G. *Spectrochim. Acta* **1984**, *39B*, 115.

58. Pollard, B. D.; Blackburn, M. B.; Nikdel, S.; Massoumi, A.; Winefordner, J. D. *Appl. Spectrosc.* **1979**, *33*, 5.

59. Epstein, M. S.; Nikdel, S.; Bradshaw, J. D.; Kosinski, M. A.; Bower, J. N.; Winefordner, J. D. *Anal. Chim. Acta* **1980**, *113*, 221.

60. Fujiwara, K.; Omenetto, N.; Bradshaw, J.; Bower, J. N.; Nikdel, S.; Winefordner, J. D. *Spectrochim. Acta* **1979**, *34B*, 317.

61. Bonczyk, P. A.; Shirley, J. A. *Combust. Flame* **1979**, *34*, 253.

62. Weeks, S. J.; Haraguchi, H.; Winefordner, J. D. *J. Quant. Spectrosc. Radiat. Transfer* **1978**, *19*, 633.

63. Haraguchi, H.; Weeks, S. J.; Winefordner, J. D. *Spectrochim. Acta* **1979**, *35A*, 391.

64. Blackburn, M. B.; Mermet, J. M.; Winefordner, J. D. *Spectrochim. Acta,* **1978**, *34A*, 847.

65. Allian, P.; Mauras, Y. *Anal. Chim. Acta* **1984**, *165*, 141.

66. Tsunoda, K.; Haraguchi, H.; Fuwa, K. *Spectrochim. Acta* **1980**, *35B*, 715.

67. Massman, H. *Talanta* **1982**, *29*, 1051.

68. Rutledge, M. J.; Tremblay, M. E.; Winefordner, J. D. *Appl. Spectrosc.* **1987**, *41*, 5.

69. Long, G. L.; Winefordner, J. D. *Anal. Chem.* **1983**, *55*, 712A.

70. Vera, J. A.; Leong, M. B.; Omenetto, N.; Smith, B. W.; Womack, B.; Winefordner, J. D. *Spectrochim. Acta* **1989**, *44B*, 939.

71. Bolshov, M. A.; Zybin, A. V.; Smirekina, I. I. *Spectrochim. Acta* **1981**, *36B*, 1143.

72. Omenetto, N.; *Spectrochim. Acta* **1989**, *44B*, 131.

73. Preli, F. R. Jr.; Dougherty, J. P.; Michel, R. G. *Anal. Chem.* **1987**, *59*, 1784.

74. Dittrich, K.; Stark, H. J.; *J. Anal. Atom. Spectrom.* **1987**, *2*, 63.

75. Kosinski, M. A.; Uchida, H.; Winefordner, J. D. *Talanta* **1983**, *30*, 339.

76. Irwin, R. L.; Wei, G. T.; Butcher, D. J.; Liang, Z.; Su, E. G.; Takahashi, J.; Walton, A. P.; Michel, R. G. *Spectrochim. Acta* **1992**, *47B*, 1497.

77. Dougherty, J. P.; Preli, F. R. Jr.; McCaffrey, J. T.; Seltzer, M. D.; Michel, R. G. *Anal. Chem.* **1987**, *59*, 1112.

78. Dougherty, J. P.; Preli, F. R. Jr.; Michel, R. G. *Talanta,* **1989**, *36*, 151.

79. Andrews, D. L. *Applied Laser Spectroscopy;* VCH: New York, 1992.

80. Ledingham, K. W. D.; Singhal, R. P. *J. Anal. Atom. Spectrom.* **1991**, *6*, 73.

81. Johnson, B. C.; Newell, V. J.; Clark, J. B.; Mcphee, E. S. *J. Opt. Soc. Am. B* **1995**, *12*, 2122.

82. Zhou, J. X.; Hou, X.; Yang, K. X.; Michel, R. G. *Abstract of Papers,* Pittcon, Chicago, IL, 1996; Abstract 591.

83. Zybin, A.; Schnürer-Patschän, C.; Niemax, K. *Spectrochim. Acta* **1992**, *47B*, 1519.

84. Bolshov, M. A.; Zybin, A. V.; Zybina, L. A.; Koloshnikov, V. G.; Majorov, I. A. *Spectrochim. Acta* **1976**, *31B*, 493.

85. Fraser, L. M.; Winefordner, J. D. *Anal. Chem.* **1973**, *44*, 1444.

86. Hecht, J.; *Laser Focus World* **1993**, May, 87.

87. E. Tashiro, Y. Oki,; Maeda, M.; Handa, C.; Hasegawa, Y.; Futami, H.; Izumi, J.; Matsuda, K. *Anal. Chem.* **1993**, *65*, 2096.

88. Green, R. B.; Travis, J. C.; Keller, R. A. *Anal. Chem.* **1976**, *48*, 1954.

89. Hecht, J. *Laser Focus World* **1993**, October, 99.

90. Rutledge, M. J.; Tremblay, M. E.; Winefordner, J. D. *Appl. Spectrosc.* **1987**, *41*, 5.

91. Duarte, F. J.; Piper, J. A. *Appl. Opt.* **1984**, *23*, 1391.

92. Womack, J. B.; Gessler, E. M.; Winefordner, J. D. *Spectrochim. Acta* **1991**, *46B*, 301.

93. Walters, P. E.; Long, G. E.; Winefordner, J. D. *Spectrochim. Acta* **1984**, *39B*, 69.

94. Seltzer, M. D.; Hendrick, M. S.; Michel, R. G. *Anal. Chem.* **1985**, *57*, 1096.

95. Omenetto, N.; Human, H. G. C.; Cavalli, P.; Rossi, G. *Analyst* **1984**, *109*, 1067.

96. Omenetto, N.; Nikdel, S.; Reeves, R. D.; Bradshaw, J. B.; Bower, J. N.; Winefordner, J. D. *Spectrochim. Acta* **1980**, *35B*, 507.

97. Smith, B. W.; Womack, J. B.; Omenetto, N.; Winefordner, J. D. *Appl. Spectrosc.* **1989**, *43*, 873.

98. Leong, M. B.; D'Silva, A. P.; Fassel, V. A. *Anal. Chem.* **1986**, *58*, 2594.

99. Funayama, M.; Mukaihara, K.; Morita, H.; Okada, T.; Tomonaga, N.; Izumi, J.; Maeda, M. *Opt. Comm.* **1993**, *102*, 457.

100. Heitmann, U.; Kötteritzsch, M.; Heitz, S.; Hese, A.; *Appl. Phys. B* **1992**, *55*, 419.

101. Andrews, D. L.; *Lasers in Chemistry;* Spring-Verlag: New York, 1986.

102. Svelto, O. *Principles of Lasers;* Plenum: New York, 1982.

103. Miyazakai, K.; Sakai, H.; Sato, T. *Opt. Lett.* **1986**, *11*, 797.

104. Shen, Y. R.; *The Principle of Non-Linear Optics;* Wiley: New York, 1984.

105. Fowles, G. R. *Introduction to Modern Optics;* Reinhart Winston: New York, 1975.

106. Massey, G. A.; Johnson, J. C. *IEEE J. Quantum Electron.* **1976**, *12*, 721.

107. Garetz, B. A.; Lombardi, J. R., Eds.; *Advanced in Laser Spectroscopy;* Wiley: New York, 1986; Vol. 3.

108. Epstein, M. S.; Bradshaw, J.; Bayer, S.; Bower, J.; Voigtman, E.; Winefordner, J. D. *Appl. Spectrosc.* **1980**, *34*, 372.

109. Kuhl, J.; Spitschan, H. *Opt. Commun.* **1973**, *7*, 256.

110. Omenetto, N.; Hatch, N. N.; Fraser, L. M.; Winefordner, J. D. *Anal. Chem.* **1973**, *45*, 195.

111. Goff, D. A.; Yeung, E. S. *Anal. Chem.* **1978**, *50*, 625.

112. Gonchakov, A.; Zorov, N. V.; Kuzyakov, Y. Y. *Zh. Anal. Khim.* **1979**, *34*, 2057.

113. Horvath, J. J.; Bradshaw, J. D.; Bower, J. N.; Epstein, M. S.; Winefordner, J. D. *Anal. Chem.* **1981**, *53*, 6.

114. Hovis, F. E.; Gelbwachs, J. A. *Anal. Chem.* **1984**, *56*, 1392.

115. Kachin, S. V.; Smith, B. W.; Winefordner, J. D. *Appl. Spectrosc.* **1985,** *39,* 587.

116. Bolshov, M. A.; Zybir, A. V.; Koloniana, L. N.; Majorov, I. A.; Smirenkina, I. I.; Shiryaeva, O. A. *Zh. Anal. Khim.* **1984,** *39,* 320.

117. Human, H. G. C.; Omenetto, N.; Cavalli, P.; Rossi, G. *Spectrochim. Acta* **1984,** *39B,* 1345.

118. Uchida, H.; Kosinski, M. A.; Winefordner, J. D. *Spectrochim. Acta* **1983,** *38B,* 5.

119. Hendrick, M. S.; Seltzer, M. D.; Michel, R. G. *Spectrosc. Lett.* **1986,** *19,* 141.

120. Smith, B. W.; Omenetto, N.; Winefordner, J. D. *Spectrochim. Acta* **1984,** *39B,* 1389.

121. Patel, B. M.; Winefordner, J. D. *Spectrochim. Acta* **1986,** *41B,* 469.

122. Dittrich, K.; Stark, H. J. *J. Anal. Atom. Spectrom.* **1986,** *1,* 237.

123. Neumann, S.; Kriese, M. *Spectrochim. Acta* **1974,** *29B,* 127.

124. Hohimer, J. P.; Hargis, P. J. Jr. *Anal. Chim. Acta* **1978,** *97,* 43.

125. Bolshov, M. A.; Zybin, A. V.; Koloshnikov, V. G.; Pisarikii, A. V.; Smirnof, A. N. *Zh. Prikl. Spectrosk.* **1978,** *28,* 45.

126. Gonchakov, A. S.; Zorov, N. B.; Kuzyakov, Y. Y.; Matveev, O. I. *Zh. Anal. Khim.* **1979,** *34,* 2312.

127. Wittman, P.; Winefordner, J. D. *Can. J. Spectrosc.* **1984,** *29,* 75.

128. Denisov, L. K.; Loshin, A. F.; Kozlov, N. A.; Nikiforov, V. G. *Zh. Prikl. Spectrosk.* **1985,** *43,* 566.

129. Simeonsson, J. B.; Ayala, N. L.; Vera, J. A.; Smith, B. W.; Winefordner, J. D. *Spectrochim. Acta* **1990,** *45B,* 1025.

130. Walters, P. E.; Barber, T. E.; Wensing, M. W.; Winefordner, J. D. *Spectrochim. Acta* **1991,** *46B,* 1015.

131. Turk, G. C.; Travis, J. C. *Spectrochim. Acta* **1990,** *45B,* 409.

132. Hueber, D.; Smith, B. W.; Madden, S.; Winefordner, J. D. *Appl. Spectrosc.* **1994,** *48,* 1213.

133. Huang, X.; Lanauze, J.; Winefordner, J. D. *Appl. Spectrosc.* **1985,** *39,* 1042.

134. Tremblay, M. E.; Smith, B. W.; Winefordner, J. D. *Anal. Chim. Acta* **1987,** *199,* 111.

135. Tremblay, M. E.; Simeonsson, J. B.; Smith, B. W.; Winefordner, J. D. *Appl. Spectrosc.* **1988,** *42,* 281.

136. Simeonsson, J. B.; Ng, K. C.; Winefordner, J. D. *Anal. Chim. Acta* **1992,** *258,* 73.

137. Omenetto, N.; Smith, B. W.; Hart, L. P.; Cavalli, P.; Rossi, G. *Spectrochim. Acta* **1985,** *40B,* 1411.

138. Greenfield, S. *J. Anal. Atom. Spectrom.* **1994,** *9,* 565.

139. Vera, J. A.; Murray, G. M.; Weeks, S. J.; Edelson, M. C. *Spectrochim. Acta* **1991,** *46B,* 1689.

140. Bolshov, M. A.; Koloshnikov, V. G.; Rudnev, S. N.; Boutron, C. F.; Görlach, U.; Patterson, C. C. *J. Anal. Atom. Spectrom.* **1992,** *7,* 99.

141. Yuzefovsky, A.I.; Lonardo, R. F.; Wang, M.; Michel, R. G. *J. Anal. Atom. Spectrom.* **1994,** *9,* 1195.

142. Hohimer, J. P.; Hargis, P. J. Jr. *Appl. Phys. Lett.* **1977,** *30,* 344.

143. Bolshov, M. A.; Rudnev, S. N. *J. Anal. Atom. Spectrom.* **1992,** *7,* 1.

144. Irwin, R. L.; Butcher, D. J.; Takahashi, J.; Wei, G.-T.; Michel, R. G. *J. Anal. Atom. Spectrom.* **1990,** *5,* 603.

145. Resto, W.; Badini, R. G.; Smith, B. W.; Stevenson, C. L.; Winefordner, J. D. *Spectrochim. Acta* **1993**, *48B*, 627.

146. Vera, J. A.; Stevenson, C. L.; Smith, B. W.; Omenetto, N.; Winefordner, J. D. *J. Anal. Atom. Spectrosc.* **1989**, *4*, 619.

147. Sjöström, S.; Axner, O.; Norberg, M. *J. Anal. Atom. Spectrom.* **1993**, *8*, 375.

148. Bolshov, M. A.; Zybin, A. V.; Koloshnikov, V. G.; Smirenkina, I. I. *Spectrochim. Acta* **1988**, *43B*, 519.

149. Bolshov, M. A.; Zybin, A. V.; Koloshnikov, V. G.; Smirenkina I.I. *Ind. Lab. (USSR)* **1989**, *55*, 1028.

150. Lonardo, R. F.; Yuzefovsky, A. I.; Irwin, R. L.; Michel, R. G. *Anal. Chem.* **1996**, *68*, 514.

151. Slavin, W. Ed.; *Graphite Furnace AAS, A Source Book;* Perkin-Elmer: Ridgefield, CT, 1984.

152. Vera, J. A.; Stevenson, C. L.; Smith, B. W.; Omenetto, N.; Winefordner, J. D. *J. Anal. Atom. Spectrosc.* **1989**, *4*, 619.

153. Remy, B.; Verhaeghe, I.; Mauchien, P. *Appl. Spectrosc.* **1990**, *10*, 1633.

154. Bolshov, M. A.; Zybin, A. V.; Smirenkina, I. I. *Spectrochim. Acta* **1981**, *36B*, 1143.

155. Bolshov, M. A.; Rudnev, S. N.; Brust, J. *Spectrochim. Acta* **1994**, *49B*, 1437.

156. Petrucci, G. A.; Beissler, H.; Matveev, O.; Cavalli, P.; Omenetto, N. *J. Anal. Atom. Spectrom.* **1995**, *10*, 885.

157. Harrison, W. W.; Bashick, C. M.; Klingler, J. A.; Ratliff, P. H.; Mei, Y. *Anal. Chem.* **1990**, *62*, 943A.

158. Broekaert, J. A. C.; *J. Anal. Atom. Spectrom.* **1987**, *2*, 537.

159. Littlejohn, D.; Jowitt, R.; Shuttler, I. L.; Sparkes, S. T.; Tyson, J. F.; Walton, S. J. *J. Anal. Atom. Spectrom.* **1990**, *5*, 179R.

160. Deavor, J. P.; Becerra, E.; Smith, B. W.; Winefordner, J. D. *Can. J. Appl. Spectrosc.* **1993**, *38*, 7.

161. Yuzefovsky, A. I.; Lonardo, R. F.; Michel, R. G. *Anal. Chem.* **1995**, *67*, 2246.

162. Omenetto, N.; Rossi, G. *Anal. Chim. Acta* **1968**, *40*, 195.

163. Omenetto, N.; *Anal. Chem.* **1976**, *48*, 75A.

164. Epstein, M. S.; Nikdel, S.; Omenetto, N.; Reeve, R.; Bradshaw, J.; Winefordner, J. D. *Anal. Chem.* **1979**, *51*, 2071.

165. Benetti, P.; Omenetto, N.; Rossi, G. *Appl. Spectrosc.* **1971**, *25*, 57.

166. Liang, Z.; Lonardo, R. F.; Michel, R. G. *Spectrochim. Acta* **1993**, *48B*, 7.

167. Farnsworth, P. B.; Smith, B. W.; Omenetto, N.; *Spectrochim. Acta* **1990**, *45B*, 1151.

168. Marunkov, A.; Chekalin, N.; Enger, J.; Axner, O. *Spectrochim. Acta* **1994**, *49B*, 1385.

169. Hatman, D. H. *Rev. Sci. Instru.* **1971**, *42*, 420.

170. Enger, J.; Malmsten, Y.; Ljiugberg, P.; Axner, O. *Analyst*, **1995**, *120*.635.

171. Preli, F. R.; Dougherty, J. P.; Michel, R. G. *Spectrochim. Acta* **1988**, *43B*, 501.

172. Klinger, D. S. *Ultrasensitive Laser Spectroscopy;* Academic: San Diego, 1983.

173. Niemax, K.; Lawrenz, J.; Obrebski, A.; Weber, K.-H. *Anal. Chem.* **1986**, *58*, 1566.

174. Lawrenz, J.; Obrebski, A.; Niemax, K. *Anal. Chem.* **1987,** *59,* 1232.

175. Obrebski, A.; Lawrenz, J.; Niemax, K.; *Spectrochim. Acta* **1990,** *45B,* 15.

176. Goldsmith, J. E. M.; Rahn, L. A. *J. Opt. Soc. Am.* B **1988,** *5,* 749.

177. Dashin, S. A.; Mayorov, I. A.; Bolshov, M. A.; *Spectrochim. Acta* **1993,** *48B,* 531.

178. Lonardo, R. F.; Yuzefovsky, A. I.; Yang, K. X.; Michel, R. G. submitted to *J. Anal. Atom. Spectrom.,* **1996,** *11,* 279.

179. Measures, R. M.; Drewall, N.; Kwong, H. S. *Phys. Rev. A.* **1977,** *16,* 1093.

180. Measures, R. M.; Kwong, H. S. *Appl. Opt.* **1979,** *18,* 281.

181. Kwong, H. S.; Measures, R. M. *Anal. Chem.* **1979,** *51,* 428.

182. Kwong, H. S.; Measures, R. M. *Appl. Opt.* **1980,** *19,* 1025.

183. Sdorra, W.; Quentmeier, A.; Niemax, K. *Mikrochim. Acta* **1986,** *II,* 201.

184. Quentmeier, A.; Sdorra, W.; Niemax, K. *Spectrochim. Acta* **1990,** *45B,* 537.

185. Moenke-Blankenburg, L. *Prog. Analyt. Spectrosc.* **1986,** *9,* 335.

186. Darke, S. A.; Tyson, J. F. *J. Anal. Atom. Spectrom.* **1993,** *8,* 145.

187. Sjöström, S.; Mauchien, P. *Spectrochim. Acta Rev.* **1993,** *15,* 153.

188. Cheam, V.; Lechner, J.; Sekera, I.; Desrosiers, R. *Anal. Chim. Acta* **1992,** *269,* 129.

189. Liang, Z.; Walton, A. P.; Butcher, D. J.; Irwin, R. L.; Michel, R. G. *Microchem. J.* **1993,** *47,* 363.

190. Enger, J.; Marunkov, A.; Chekalin, N.; Axner, O. *J. Anal. Atom. Spectrom.* **1995,** *10,* 539.

191. Butcher, D. J.; Irwin, R. L.; Takahashi, J.; Su, E. G.; Wei, G.-T.; Michel, R. G. *Appl. Spectrosc.* **1990,** *44,* 1521.

192. Dittrich, K.; Hanisch, B.; Stärk, H-J. *Fresenius Z. Anal. Chem.* **1986,** *324,* 497.

193. Anwar, J.; Anzano, J. M.; Petrucci, G.; Winfordner, J. D. *Analyst* **1991,** *116,* 1025.

194. Anwar, J.; Anzano, J. M.; Petrucci, G.; Winfordner, J. D. *Michrochem. J.* **1991,** *43,* 77.

195. Anwar, J.; Anzano, J. M.; Winfordner, J. D. *Talanta* **1991,** *38,* 1071.

196. Su, E. G.; Michel, R. G. *J. Anal. Atom. Spectrom.* **1994,** *9,* 501.

197. Butcher, D. J.; Irwin, R. L.; Takahashi, J.; Michel, R. G. *J. Anal. Atom. Spectrom.* **1991,** *6,* 9.

198. Yuzefovsky, A. I.; Michel, R. G. *J. Anal. Atom. Spectrom.* **1994,** *9,* 1203.

199. Winefordner, J. D.; Smith, B. W.; Omenetto, N. *Spectrochim. Acta* **1989,** *44B,* 1397.

200. Hieftje, G. M.; *J. Chem. Ed.* **1982,** *59,* 900.

201. Frech, W.; Baxter, D. C.; *Spectrochim. Acta* **1990,** *45B,* 867.

202. L'Vov, B. V. *Spectrochim. Acta* **1990,** *45B,* 633.

203. Omenetto, N. *Mikrochimica Acta* **1991,** *II,* 277.

204. Hill, S. J.; Dawson, J. B.; Price, W. J.; Shuttler, I. L.; Tyson, J. F.; *J. Anal. Atom. Spectrom.* **1995,** *10,* 224R.

# Laser Ablation for Sample Introduction: Principles and Applications

*Lieselotte Moenke-Blankenburg*

## 4.1 Introduction

Among known techniques of direct solid sampling, laser ablation offers a significant potential for the in situ analysis of solids in micro and macro regions. Direct solid sampling by laser ablation (LA) is an attractive method for sample introduction to atomic emission spectrometry (AES) and atomic mass spectrometry (MS) since it can avoid the time-consuming sample digestion and preparation procedures usually required for most of the sample introduction techniques.

The direct analysis of solids can be carried out without the use of chemical reagents and without processes of separation and concentration which could introduce contamination or could cause losses of the analyte. Solid sampling techniques have additional advantages when the sample is difficult or hazardous to digest. For instance, certain compounds of refractory carbides and oxides, silicate minerals, and others do not readily digest, dissolve, or remain in solution. Dissolution will also dilute the often already low concentrations of the analytical elements, lowering the power of detection. Solution nebulization delivers only a small percentage of the sample into the plasma, whereas direct solid injection sends most of the sample into the excitation or ionization source.

Since the 1960s, soon after the first report of laser action in ruby,[1] it was generally recognized that the intense laser output beam could be used to excite material into a state of optical emission[2-12] and to ionize material suitable for mass separation.[13-27] From that time to the present, the field of laser application in atomic spectroscopy has been evolving and changing. Up to now, efforts have been concentrated on the development of special laser types and laser parameters, and the improvement of quantification and reproducibility in direct techniques, for example, laser ablation atomic emission spectrometry (LA-AES), but especially in tandem techniques such as laser ablation inductively coupled plasma atomic emission spectrometry (LA-ICP-AES)[28] and laser ablation inductively coupled plasma mass spectrometry (LA-ICP-MS).[28]

The principle of solid sampling by laser ablation is as follows:

- A solid specimen of an appropriate size is set up on an $x$-$y$-$z$ movable stage which may be placed in a specially designed laser ablation chamber.
- The surface of the specimen is investigated by an optical lens or under a microscope and the region to be analyzed chemically is selected.
- The parallel radiation of a solid-state laser, gas laser, or excimer laser with suitable output energy, time duration, and therefore, output power is focused by a lens or an optical system onto a small spot of a selected region of the sample.
- The nanogram to milligram amount of ablated sample material, produced by one or more laser shots, consists of small solid particles, solidified liquid droplets, clusters, free atoms, and more or less atomized, excited, and ionized vapor (Figures 4.1 and 4.2). If the vapor contains significant populations of excited and/or ionized atoms, direct LA-AES or direct LA-MS for elemental analysis is possible.

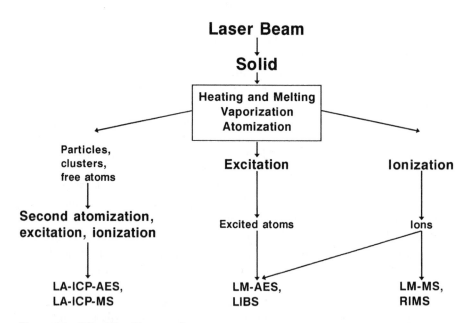

**Figure 4.1** Principle of laser action.

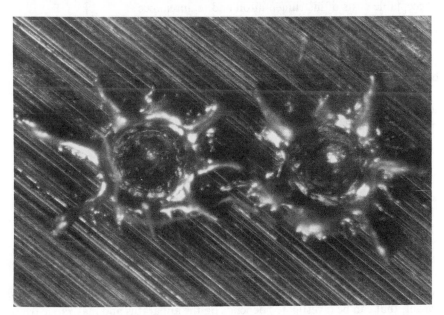

**Figure 4.2** **(a)** Effect of a focused laser beam with an output energy of 500 mJ (normal mode) and a wavelength of 1064 nm on a steel sample: besides the luminous sample vapor cloud being emitted, solid particles are ejected from the surface of the sample and reflected on the front lens of the objective. **(b)** Two craters obtained after single shots of free-running IR-laser radiation (1064 nm) with an output energy of 500 mJ focused on the surface of a steel sample: craters surrounded by splashed substance.

- The ablated material can be introduced into an ICP as an aerosol by the flow of a carrier gas, for instance, argon. With the high temperature of the ICP, the ablated material will be, in a second step, completely atomized, excited and, if wanted, ionized. The radiation of the atomic emission or the masses of isotopes can be recorded with different kinds of spectrometers of LA-ICP-AES and LA-ICP-MS.

An ideal solid sampling method should include the following features:

- Simultaneous multi-element analysis
- Applicable to a wide range of sample composition
- Wide dynamic range of concentrations to be determined
- Handle small and large sample sizes
- Simple changing of samples
- Simple operation
- Reproducible sampling
- Accurate measurement
- Variability from micro-analysis (local) to distribution analysis (profiles lateral and vertical as well as areas) and to macro-analysis (bulk)
- Suitable methods for quantification
- Acceptable costs of instrumentation and maintenance

Despite the increased performance of laser ablation techniques, all these techniques and their available approaches are awaiting adoption, improvement, evaluation, optimization, and application in many fields of research and quality control.

Laser ablation has four main fields of application: It can be a microchemical method, a method for local analysis in microregions, a method for distribution analysis with spatial resolution in microregions, and a method of bulk analysis.

- Microchemical analysis of nonconducting as well as conducting powders with laser ablation requires, in principle, very small amounts of substance (about 0.1 µg to 0.1 mg). Powders of sufficient quantity have to be applied to an object disk with or without an adhesive or have to be embedded.
- Local analysis in microregions offers a spatial resolving power of about 1–100 µm. The method also provides information on the change in the composition of the specimen as a function of the spatial coordinates; thus distribution analysis is possible.
- A line analysis is achieved by a sequence of local analyses with a regular stepwise movement of the specimen carriage after each laser shot. The length of the profile that can be investigated depends on the apparatus and may range from ten micrometers to several millimeters. A depth profile is performed by subsequent focusing of the laser radiation step by step from 1 to about 100 µm.
- An area analysis is accomplished by a successive sequence of spot analysis in two dimensions. An area of about 10–100 mm$^2$ can be recorded.
- A layer analysis is attained by a vertical sequence of area analyses after stepwise focusing of the laser radiation in depth from 1 to about 100 µm.

- A bulk analysis could be brought about by an effort of summation of single shots. A volume of several cubic millimeters can be investigated.

Atomic emission spectrometry and mass spectrometry permit about 60 chemical elements to be detected simultaneously in AES and sequentially (with high-speed scanning instruments) in MS, as well as simultaneously by different methods of separation and detection. In addition, isotope analysis can be achieved after mass separation in MS.

## 4.2  Laser Types Suitable for Ablation of Solid Samples

The well-known properties of laser beams are high intensity, directionality, and monochromaticity. The main attractive feature of laser ablation is the ability to sample, vaporize, atomize, excite, and ionize both conducting and nonconducting solids in micro- and macroregions.

The most frequently used lasers are solid-state lasers, such as ruby (694 nm) and Nd:YAG (1064 nm, Q-switched (pulse length 5–10 ns), frequency-doubled (532 nm, pulse length 5–8 ns), frequency-tripled (355 nm, pulse length 4–8 ns), frequency-quadrupled (266 nm, pulse length 3–7 ns), and also gas lasers, such as $CO_2$ lasers (10.6 μm) and $N_2$ lasers (337 nm). In the future, more and more applications are to be expected for excimer lasers, such as XeF (351 nm, pulse length 20 ns), XeCl (308 nm, pulse length 20–45 ns), KrF (248 nm, pulse length 20–40 ns), KrCl (222 nm, pulse length 20–40 ns), and ArF lasers (193 nm, pulse length 20–40 ns).

The ruby laser is an example of an ionic crystal laser which uses a crystal of $Al_2O_3$ doped with 0.05% by mass $Cr_2O_3$, where the chromium ions ($Cr^{3+}$) substitute the aluminum ions ($Al^{3+}$). The chromium ion is the active species and has an absorption band with wavelengths from about 500 to 600 nm that can be used by an optical pump. The ruby represents a typical three-level system. The active material, formed by a single ruby crystal, is shaped as a cylindrical rod with a diameter of a few millimeters and a length of a few centimeters. Since the bandwidths of the pump transitions are quite large (about 50 nm), optical pumping can be used conveniently. The ruby laser normally operates in a pulsed regime, and the pumping is accomplished by flashlamps such as low-pressure Xe lamps, although it can also operate in a continuous wave (CW) regime (when pumped by high-pressure Hg capillary lamps).

The ions that are excited by the pump decay to a fluorescing energy level that has a lifetime of 3 ms and causes amplification at 694.3 nm. Millisecond trains of many submicrosecond pulses are obtained when millisecond flashlamp pump pulses are used and the cavity is not Q-switched.

The pulsed ruby laser was the first laser to be operated (in 1960) and was widely used to the eighties[29–37] and nineties.[38–47]

A neodymium laser is a common solid-state ion laser that uses the $Nd^{3+}$ ion in a suitable glass or host crystal such as yttrium aluminum garnet (YAG), Y-

$_3Al_5O_{12}$. The absorption bands for pumping the $Nd^{3+}$ lasers are in the red region of the spectrum, while the wavelength that is amplified is in the near infrared, 1064 nm, with a lifetime of about 0.5 ms.

The energy levels involved form a four-level system, which facilitates continuous amplification and lasing with a gas discharge lamp pump. As for the ruby laser, the active material is shaped as a cylindrical rod and is optically pumped. The Nd:YAG laser can operate both in a pulsed regime (with low-pressure Xe flashlamps as the excitation source) or in the CW mode (with W or Kr pump lamps). Now, the Nd:YAG laser is mostly used in LA-ICP-MS[48–76] and LA-ICP-AES.[44,46,47,77–82] Frequency doubling at 532 nm,[78,83–89] frequency tripling,[83,84,87,88,90] and frequency quadrupling at 266 nm[83,85,91–95] is more and more often included with these tandem techniques.

Gases have narrow absorption bands or lines. Therefore, optical pumping with flashlamps is not efficient for gas lasers except in very special cases where the pump has narrow lines that happen to overlap the narrow absorption peaks. Electrical discharges are therefore commonly used to pump the atoms, ions, or molecules in a gas optical amplifier.

The carbon dioxide laser utilizes rotovibrational transitions of the ground electronic state of the $CO_2$ molecule. This molecule, which is triatomic and linear, possesses three nondegenerate vibration modes: the longitudinal symmetric mode, the transverse (or bending) mode, and the longitudinal antisymmetric mode. The laser action usually occurs on a series of rotational lines with the wavelength of 10.6 μm. It is also possible to get laser action on some rotational lines of a second transition with the wavelength of 9.6 μm. The upper laser level can be pumped by collision of the $CO_2$ molecules with electrons or, more efficiently, by resonant energy transfer from $N_2$ molecules that are purposely added to the gas. The gas mixture also contains He, which reduces the lifetime of the lower levels through collisions and also contributes to the cooling of the gas. The $CO_2$ molecular laser has a high power efficiency (up to 30%) and high power in the infrared region of the spectrum at 10.6 μm. Pulsed and continuous wave operation are common.[96,97]

The $N_2$ laser uses vibrational transitions of the molecules of nitrogen. The emission wavelength (337.1 nm) falls within the ultraviolet. The laser is self-terminating since the lifetime of the upper level is shorter than that of the lower level; therefore, the $N_2$ laser can be operated only in a pulsed regime. The laser action requires a very short, powerful pump current pulse. This can easily be achieved with a transverse discharge and fast electronic circuit. A current pulse of duration as short as 1 ns and a peak value of up to $10^5$A can be obtained and used. Nanosecond laser pulses with a power of several hundred kilowatts are reachable.[98]

Excimer (excited-dimer) lasers are based on molecular systems that have bound excited states but no bound ground states. Examples of these systems are rare-gas halide molecules, such as XeF, XeCl, KrF, KrCl, and ArF, which have excited ion-pair states ($A^+B^-$), where the rare-gas atom A forms a positive ion, and the halogen atom B easily attaches an electron to form a negative ion.

These ions attract each other on a Coulomb potential. If the electron on the B ion radiatively recombines with the A ion, the molecule will fall to the neutral AB ground-state potential (Figure 4.3). There is no Coulomb attraction here, so the core electron repulsion causes a rapid dissociation of the molecule, and there is automatically a population inversion on the transition between the excited and ground states. The excimer lasers use a fast (20–50 ns) electrical discharge pulse to excite the upper laser level. Recent experimentation is demonstrating that excimers are becoming more and more popular because of their

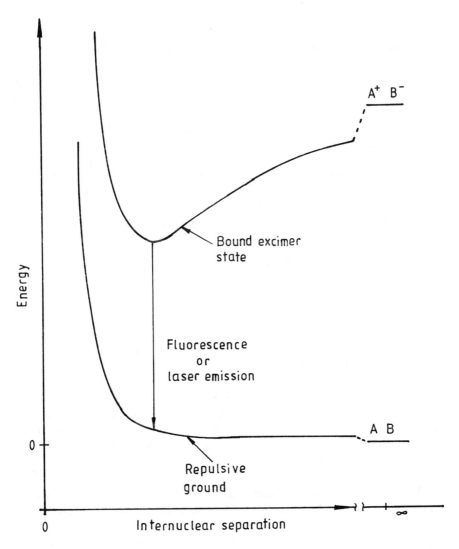

**Figure 4.3** Scheme of the energy levels of an excimer molecule, such as a rare-gas halide.

ultraviolet wavelengths, short pulse duration, suitable peak power, and variable repetition rate.[82,99-106]

When applying lasers in analytical chemistry, one must ensure that the chosen laser meets all of the many requirements: wavelength, mode (CW or pulsed, normal or free-running, fixed Q or Q-switched), output energy, energy density (fluence), output power (radiant flux), power density (irradiance), time duration (pulse time, rise time), beam profile (energy distribution, divergence), and repetition rate.

Examples of suitable parameters of different lasers for the use in elemental spectrochemistry are given in the Tables 4.1–4.5.

## 4.3  Interactions and Transactions in Laser Ablation

Some of the present limitations on the performance of LA-ICP-spectrometry arise from a lack of information concerning the processes by which material is mobilized from a solid surface as a result of the interaction of laser radiation with the solid surface. The interaction of laser light with solids, and subsequent development of an expanding radiant plume, has been reviewed in detail by several authors.[107-122] Diagnoses are based on the parameters of plumes, which are not in equilibrium, as supposed, due to the presence of an external energy source (in the present case, a radiation source). It is well known that nonequilibrium phenomena are among the most complicated problems to investigate because of the difficulty of defining a unique temperature for the whole system.

### 4.3.1  Focusing of Laser Radiation

One attribute of laser radiation that is of interest as it affects ablation of solids is the parallelism of the beam. Typically, for suitable lasers in laser microanalysis, the spreading angle is on the order of a few milliradians. Therefore, it is easy to focus all the radiation on a small spot by means of an optical lens system. The spot diameter depends on the wavelength and the divergence angle of the laser and on the focal length and the aperture of the focusing optics. A laser beam with well-defined modes, by preference a single mode, can be focused to a very sharp spot of a minimum size. When such a single mode beam is focused with an aberration-free lens of focal length $f$ and radius $r$, a Gaussian distribution of power density is obtained at the focal plane. The radius of the focal spot $r_s$ is directly proportional to the focal length as shown in the equation

$$r_s \approx \lambda f / \pi r. \tag{4.1}$$

When a Gaussian beam is passed through a circular aperture, the resulting beam diverges with an Airy function, which has a central maximum surrounded by circular fringes. If the first fringe is used to define the radius of the focal spot, then the spot has a radius $r_s$ equal to $1.22 \, \lambda \, f/2r$ and equal to $\Theta f$, where $\Theta$ is the divergence angle of the laser beam in radians.

**Table 4.1** Parameters of Ruby Lasers in Use for ICP Spectrometry

| Laser type | Wavelength (μm) | Mode, repetition rate | Output energy (J) | Output power (W) | Time duration (μs) | Power density (W cm$^{-2}$) | References |
|---|---|---|---|---|---|---|---|
| LMA-10[a] | 0.694 | normal | 1 | $10^4$ | ≤1000 | $10^7$–$10^{10}$ | 29, 32–47 |
| | | semi-Q | 0.9–0.2 | ~$10^5$ | | | 29, 32–47 |
| Korad-K1[b] | 0.694 | Q, 4 min$^{-1}$ | 0.1 | $2 \cdot 10^6$ | 0.05 | $10^9$–$10^{12}$ | 29, 32–47 |
| | | normal | 1–2 | | ~1000 | | 30 |
| | | Q, 15 min$^{-1}$ | | $10^8$ | 0.01–0.02 | | 30 |
| JLR-02A[c] | 0.694 | normal | 30, 18 | | | | 31 |
| | | Q, 1 min$^{-1}$ | 3, 2 | | | | 31 |

[a] Laser Micro Analyzer LMA-10, Carl Zeiss Jena, Germany
[b] Korad-K1 with KDP Pockels cells 212-150, Interactive Radiation Inc.
[c] JEOL JLR-02A

**Table 4.2** Range of Parameters of Q-Switched Nd:YAG Lasers Used in ICP Spectrometry, 1989–1995

| | |
|---|---|
| Wavelengths | 1064, 532, 355, 266 nm |
| Output energy | 1–350 mJ |
| Output power | $10^5$–$10^8$ W |
| Mean power density | $10^9$–$10^{12}$ W cm$^{-2}$ |
| Pulse time duration | 2.5–16 ns |
| Pulse repetition rate | 1–20 Hz |

**Table 4.3** Parameters of Commercially Available Nd:YAG Lasers

| | Perkin-Elmer 320 Laser Sampler[a] | VG, Fisons Laserlab | Finnigan Mat Laser Ablation System 266 | Cetac Technologies Inc./ FMS Analyt. Consult. LSX-100 Laser Solid Sampler | | |
|---|---|---|---|---|---|---|
| Wavelengths (nm) | 1064 | 1064 | 266 | 1064 | 532 | 266 |
| Output energy (J) (free-running) | 0.5 | 0.5 | 0.05 | | | |
| Pulse chain length (μs) | 140 | 140 | | | | |
| Output energy (J) (Q-switched) | 0.320 | 0.250 | 0.030 | *TEM*[00] 0.250 0.025 | 0.100 0.010 | 0.030 0.003 |
| Single pulse length (ns) | 2.5; 8 | 8–10 | 8–14 | . . . . . . . 8–10 . . . . . . . | | |
| Pulse repetition rate (Hz) | 1–15 | 1–15 | 1–5 | . . . . . . . 1–20 . . . . . . . | | |
| Range of spot sizes (μm) | 10–6000 | 10–300 | 20–200 | . . . . . . . 5–200 . . . . . . | | |

[a] Delivered until 1995

**Table 4.4** Parameters of Excimer Lasers Used in ICP Spectrometry

| Laser type | Wavelength (nm) | Repetition rate (Hz) | Output energy (mJ) | Time duration (ns) | Power density (W cm$^{-2}$) | References |
|---|---|---|---|---|---|---|
| XeF | 351 | | | 20 | | 100 |
| XeCl | 308 | | | 28 | | 101 |
| XeCl | 308 | | 250 | 20 | | 82 |
| XeCl | 308 | | 150 | 28 | $1.7 \times 10^9$ | 102 |
| KrF | 248 | | 20–300 | 15 | | 99 |
| KrF | 248 | 1–20 | 100 | 12 | | 103 |
| KrF | 248 | | 100 | | $1.5 \times 10^9$ | 104 |
| KrF | 248 | 10 | 30 | 30 | | 105 |
| KrF | 248 | 5 | | 30 | $0.2$–$40 \times 10^9$ | 106 |

**Table 4.5** Parameters of Commercially Available Excimer Lasers Suitable for ICP Spectrometry

| Laser | Wavelength (nm) | Max. repetition rate (Hz) | Output energy (mJ) | Time duration (ns) | Average power (W) | Manufacturer and type |
|---|---|---|---|---|---|---|
| XeF | 351 | 20 | 150 | | 3 | Lambda Physik: Compex 102, |
| XeF | 351 | 100 | 150 | | 12 | Compex 110, |
| XeF | 351 | 200 | 80 | | 12 | Compex 120 |
| XeF | 351 | 100, 200 | 100 | 20 | 10, 20 | LPX 110i, 120i |
| XeCl | 308 | 5 | 300 (at 1 Hz) | 40 | 1 (at 5 Hz) | Sopra: SEL 520, |
| XeCl | 308 | 10 | 250 (at 10 Hz) | 40 | 2.5 (at 10 Hz) | SEL 521 |
| XeCl | 308 | 20 | 200 | | 4 | Lambda Physik: Compex 102, |
| XeCl | 308 | 100 | 200 | | 16 | Compex 110, |
| XeCl | 308 | 200 | 100 | | 20 | Compex 120, |
| XeCl | 308 | 100, 200 | 200 | 17 | 16, 32 | LPX 110i, 120i |
| KrF | 248 | 3 | 400 (at 1 Hz) | 40 | | Sopra: SEL 520 |
| KrF | 248 | 3 (10) | 250 (at 10 Hz) | 40 | | SEL 521 |
| KrF | 248 | 20 | 350 | | 7 | Lambda Physik: Compex 102, |
| KrF | 248 | 100 | 350 | | 25 | Compex 110, |
| KrF | 248 | 200 | 200 | | 30 | Compex 120 |
| KrF | 248 | 100, 200 | 300 | 23 | 30, 60 | LPX110i, 120i |
| ArF | 193 | 2 | 200 (at 1 Hz) | 15 | 0.2 | Sopra: SEL 515 |
| ArF | 193 | 5 | 150 (at 5 Hz) | 15 | 0.75 | SEL 516 |
| ArF | 193 | 20 | 200 | | 4 | Lambda Physik: Compex 102 |
| ArF | 193 | 100 | 200 | | 12 | Compex 110 |
| Arf | 193 | 150 | 100 | | 12 | Compex 120 |
| ArF | 193 | 100, 200 | 200 | 17 | 13, 26 | LPX110i, 120i |

### 4.3.2 Ablation and Vaporization of Solids

Ready[109] regards the optical energy as being turned into heat instantaneously at the point at which the light was absorbed. The intense local heating experienced by the target causes a rapid rise in the surface temperature of the material. Heat is conducted into the interior of the target and a thin molten layer forms below the surface. As the thermal energy deposited at the surface increases, a point is reached where the deposited energy exceeds the latent heat of vaporization. When this happens, heat cannot be conducted away from the point of irradiation fast enough to prevent the surface from reaching its boiling temperature, and evaporation occurs from the surface. The distribution occurs so rapidly in the time scale of Q-switched and free-running laser pulses that Ready[109] assumes a local equilibrium to have been established during the pulse. This assumption may break down for the case of very short pulses.

The energy density deposited at the surface of the target by a power density $F$ is $Ft_e$; hence, the average energy per unit mass acquired by the thin layer of molten metal is $Ft_e/d(at_e)^{1/2}$, where $d$ is the mass density of the target, $a$ the thermal diffusivity, and $t_e$ the duration of the laser pulse. For evaporation to occur, the energy deposited in this layer must exceed the latent heat of vaporization of the target, $L_V$. Thus, the following threshold condition is obtained for the minimum absorbed power density ($F_{min}$), below which no evaporation will occur:

$$F_{min} = dL_V a^{1/2} t_e^{-1/2} \qquad (4.2a)$$

The following is a more modern expression:

$$F_V = \rho Q \sqrt{a/t_L} \qquad (4.2b)$$

| | | | |
|---|---|---|---|
| $F$ | = | power density | (W/m²) |
| $v$ | = | vaporization | |
| $\rho$ | = | density | (kg/m³) |
| $Q$ | = | specific thermal amount | (Ws/kg) |
| $a$ | = | thermal conductivity | (m²/s) |
| $t_L$ | = | time duration per laser pulse | (s) |

### 4.3.3 Ablation and Vaporization by *CW* Lasing, Free-Running, and Semi-Q-Switched Laser Pulses

Comparing the interactions with a target by a continuous wave laser, a free-running (normal, fixed Q) laser pulse chain, and either a semi-Q-switched laser with a variable number of spikes or a Q-switched laser giant pulse with or without repetition of the pulses, it can be stated that there is a great difference in the behavior of surfaces struck by these lasers.

Lasers of the first category are typified by higher energies, longer pulse or pulse chain lengths, and lower powers and power densities, as opposed to lasers of the second category, which have lower energies, much shorter pulse lengths, and therefore, higher powers and power densities.

A low-power laser with continuous or long pulse chain duration, when focused, raises the surface of the material to the vaporization temperature (boiling point) in a short time, depending on the thermal conductivity, heat capacity per unit mass, mass density, and the vaporization temperature of the sample. As more energy is delivered to the surface, the initial vaporization process changes to a melting-spraying mechanism, by which a significant amount of material is removed from the crater. This process may occur several times during the same laser shot. The stream of the aerosol leaves the surface with a velocity on the order of about $10^4$–$10^9$ cm/s.

### 4.3.4 Ablation and Vaporization by Q-Switched Giant Pulse Lasers

A higher-power laser with short pulse time duration, when focused, causes considerable ablation pressure on the target. The mechanical effects of shock waves can cause fragmentation of the sample. The temperature of the vapor leaving the surface is higher than the boiling temperature and the vapor is therefore partly ionized. The microplume formed above the surface can absorb the later part of the incoming laser spike (Figure 4.4).

In contrast to the small, deep crater caused by a free-running laser (Figure 4.5), the result of the interaction of a giant pulse laser with a target is a shallow crater with a larger diameter (Figure 4.6). A typical nonequilibrium laser plume

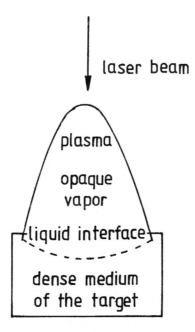

**Figure 4.4**  Assumption of laser target interaction.

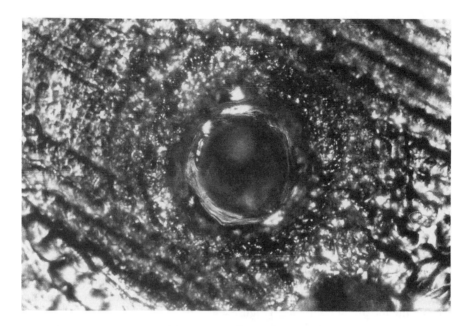

**Figure 4.5**    Crater 10 μm in diameter and 10 μm in depth produced by one free-running shot.

generated by pulsed laser radiation with a power density of $10^9$ W/cm$^2$ is characterized by the physical parameters given by Mazhukin.[120] Under a power density of about $10^9$ W/cm$^2$, the nonequilibrium in the plume is a result of its significant overheating. The concentration of electrons, ions, and atoms in ground and excited states exceeds the Saha-Boltzmann distribution.

### 4.3.5  Influences of Sample Parameters on Ablation and Vaporization

Ablation is affected by the absorptivity of the samples. Material with high absorptivity of a well-chosen wavelength ablate well, forming cylindrical craters with circular-shaped, flat, clean edges. With transparent materials the ablation is weak, and defects or inclusions in the sample will create irregularly shaped pits. Because many materials have higher absorptivities at lower wavelengths (e.g., Nd:YAG at 266 nm, and excimers) the characteristic of ablation can essentially be improved. Moreover, the smaller focal spot obtained at shorter wavelengths permits correction of the spatial resolution. Crater diameters of less than 10 μm can be achieved. The IR wavelength (e.g., Nd:YAG at 1064 nm) and the energy range of commercially available laser samplers is not well suited to biological materials. The surface of the sample is burned down and

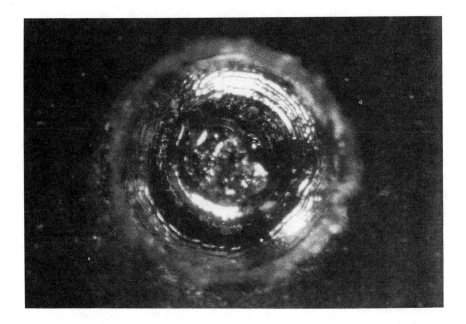

**Figure 4.6**  Crater 20 μm in diameter and a few microns in depth produced by a Q-switched single shot.

large spot sizes result. It is difficult to achieve a spot diameter of less than 50 μm. The use of shorter wavelengths and control of the energy is highly recommend. Very often, UV ablation is a better choice for analytical purposes because it reduce matrix effects, whereas selective fractionation and matrix effects are known results of IR and visible laser ablation. Also, UV lasers produce more transparent plumes with higher rates of excited and ionized species.

Other physical properties of the sample to be analyzed (Figure 4.7), such as reflection, scattering, transmission, absorption, thermal conductivity, and so on influence the process of efficaciousness of laser radiation at surfaces.[109,117] In addition, consideration of energy losses and, therefore, energy balance measurements are important (see reference 117, pp. 43–48).

All attempts to provide theoretical explanations suffer from the complexity of the phenomena. Nevertheless, Vertes et al.[123–126] developed a one-component one-dimensional hydrodynamic model to describe the expansion of laser-generated plumes. A comprehensive study of the model covered three different laser types (ruby, $CO_2$, and frequency-quadrupled Nd:YAG) and three classes of solid targets (metals, transparent insulators, and opaque insulators). The full characterization of plumes is problematical but approximate methods for diagnostics and modeling have been used to obtain a better understanding of the processes involved and to increase the reliability of the analytical results.[126–129]

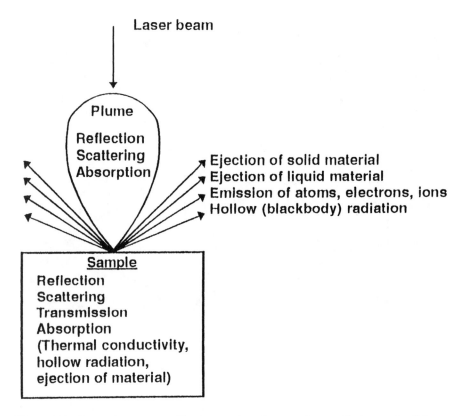

**Figure 4.7** Energy losses caused by interactions between laser radiation and solid surfaces[117] (pp. 43–48); reprinted with permission of Wiley & Sons, New York.

## 4.4 Laser Ablation Atomic Emission Spectrometry and Laser Ablation Inductively Coupled Plasma Atomic Emission Spectrometry

### 4.4.1 Principle of LA-AES

The technique of direct spectrochemical analysis by using laser plasma excitation has been known since 1962[2,3] and was developed as a microanalytical method in the early sixties[4-12] (see review 130 and 131). Development continued during the seventies (see reviews in references 117, 118, and 132) and eighties (see reviews in references 112–115, 117, 118). The method is still well established in analytical chemistry.[40,45,82,133-145] It is a simple method because it uses the single-step ability of focused laser radiation to vaporize *and* excite material, especially solids. The spectral emission of the laser-induced plume is recorded in order to determine and quantify the elemental composition of an unknown sample. The principles are described in Figure 4.8[102] and Figure 4.9.[82] The prin-

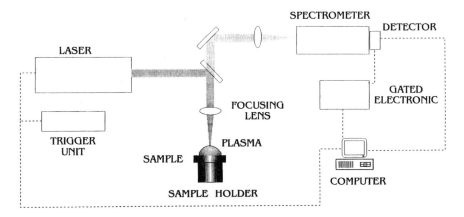

**Figure 4.8** Experimental arrangement[102] of LA-AES; reproduced with permission of Elsevier Science, Amsterdam.

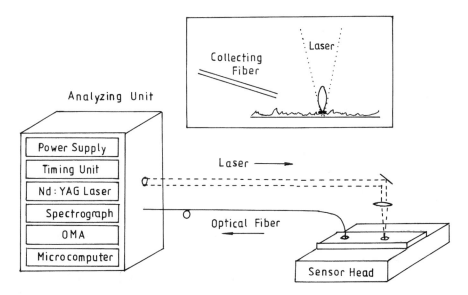

**Figure 4.9** Experimental setup[82] of LA-AES. The analyzing unit is separated from the detector head. For some applications, the laser light will be transferred through an optical fiber; thus the small detector head will be easily manipulated. The inset shows schematically the laser spark and the collecting fiber positioning. The detector head consists of the focusing lens, collecting fiber (adjusted by a mechanical manipulator), and sample holder. The sample is continuously rotated and its surface is flatted. Reprinted with permission of the American Chemical Society.

ciples of laser ablation atomic emission spectrometry (LA-AES), which some authors[121,122] have recently called laser-induced breakdown spectrometry (LIBS), are shown in Figure 4.10.[121] This is a technique that offers several very attractive features, especially when, as is usually the case, it is performed at atmospheric pressure in air. Advantages are that the analysis of solids requires little or no sample preparation, there are no restrictions in sample sizes, and measurements can be accomplished at a distance, without any contact between the analytical equipment and the analytical sample. That is important for industrial applications, as in on-line quality process control.

When comparing the current results of LA-AES to those of the sixties and seventies, especially where low reproducibility of results is concerned, the progress in laser techniques, spectrometer systems, and data treatment electronics shows that the earlier problems were not fundamentally insuperable. Reproducibility of a few percent on measurements can now be obtained.

### 4.4.2 Instrumentation of LA-AES

#### 4.4.2.1 Laser Microscopes

The oldest form of laser microprobe was developed in 1962.[3] For any true microspectrochemical analysis a microscope and a dispersing instrument are needed. A microscope fitted for visual observation, measuring, and irradiation of very small sample zones with time- and space-coherent laser radiation, linked

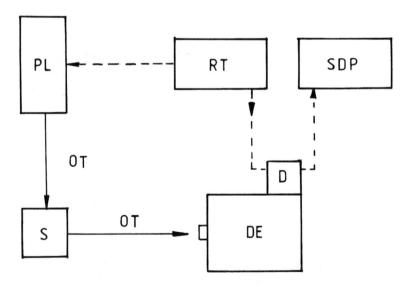

**Figure 4.10** A generic laser-induced plasma experiment (LIBS).[121] PL, pulsed laser; OT, optical train (lenses or fiber optic); S, sample; DE, dispersing element (spectrometer); D, detector (detector array or PMTs); RT, master timer; SDP, signal data processing. Reprinted with permission of Academic Press, San Diego, CA.

with a spectrograph provided the first really useful and practical outfit for microspectrochemical emission analysis for wavelengths from 200 to 1000 nm. The first laser microanalyzers, LMA and LMA1 made by Carl Zeiss Jena, Germany (see reference 131, Figure 3, p. 20), and Mark I and II made by Jarrell-Ash, Waltham, MA, USA (see reference 131 Figure 4, p. 22, 23), were commercially available in 1964/65. An essential feature of a laser microscope is that the objectives for both the observation of the specimen and for focusing the laser radiation must be suitable for the ablation, vaporization, and excitation of the material. The design of such a microscope and the path of the beams are shown in Figure 4.11. As shown in this figure, a microscope consists of a base containing the optical system for the transmitted light, a transverse carrier for the second part of the optical pathway of the reflected beam, the optical system for the pathway of the laser beam and the autocollimator for adjusting the laser, and another optical system for the observation beam and photomicrography. On the left of the vertical carrier is the laser box between the base and the transverse carrier. It contains solid-state lasers, such as ruby, Nd:glass or

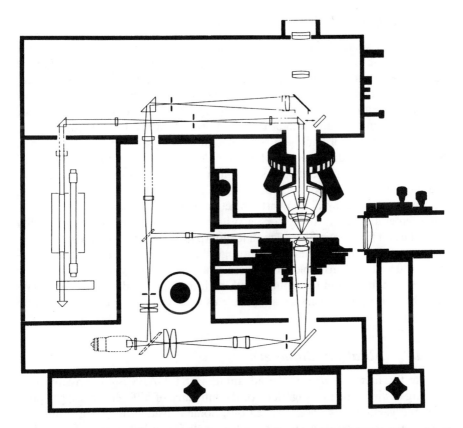

**Figure 4.11** Optical pathway in a laser microscope for LA-AES (LMA10).[117] Reprinted with permission of Wiley & Sons, New York.

Nd:YAG as active media and a flashtube as the optical pump, arranged at the foci of an elliptical reflector, and also contains the Q-switch device. On the right side of the microscope is the sample stage and the turret head with the objectives. In front of the opening on the side turned toward the spectrometer is a condenser that images the radiation emitted from the cloud of vapor of the material under investigation onto spectral apparatus.

Laser microscopes need highly specialized objectives (Figure 4.12) when using lasers with output energies above 1J per pulse or with power in the kilowatt to megawatt range. Laser radiation with higher performances may damage cemented optical components, coatings, and mirror surfaces. A compromise in working distances is required, taking into account the desired focal length (and, therefore, the diameter of the focus), the technical arrangement of the sample carrier and sample environment, and the protection of the front lens against contamination by sample vapor and sputtered particles. Parameters for microscope objectives and simple lenses used only for focusing the laser beam are shown in Table 4.6.

### 4.4.2.2 Optical Fibers

Optical fibers capable of suitable transmission were, in certain cases, placed near the laser plume or beneath additional optical components to transfer the emitted light to the entrance list of the spectrometer.[82,146,147] Wisbrun et al.[82] used a bundle of ten optical fibers made from quartz and silica, each 200 μm in inner diameter and about 20 m long, for the purpose of developing a detector for heavy metals in soil, sand, and other soft industrial wastes. Jowitt[146] investigated principles aimed at providing in situ liquid steel analysis. Various light guides, ranging from 1-cm diameter multifiber bundles to seven 0.5-mm fibers arranged in a slit formation at one end, to single fibers of various diameters were tested. The most efficient system was found to be a single fiber made from vacuum deposit quartz in conjunction with objective and imaging lenses (Figure 4.13). Lorenzen et al.[147] tried to develope a system for industrial process and quality control using fiber optic fibers 3 m in length (parameters not specified in detail). They concluded that transmission through the fiber near the sulfur triplet (between 180 and 186 nm) rapidly decreases. Furthermore, they concluded that the transfer of nanosecond laser pulses with high peak power densities ($>0.1$ GW/cm$^2$) through optical fibers causes problems even when core diameters are up to 2 mm.

### 4.4.2.3 Ablation Chambers

Despite the fact that one of the advantages of LA-AES is the possibility of working in air, some investigations of the influence of different gases, gas pressure, and vacuum have been carried out. Some considerations concerning the construction of vacuum and gas chambers include the following: (1) extending the spectral range into the far-UV field, particularly to perform C, P, and S

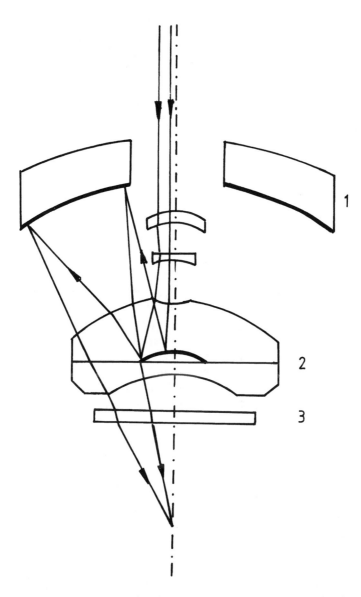

**Figure 4.12** Ray diagram of microscope mirror-lens objective, a reflecting flat-field apochromat ×40/0.5 ∞/0 (part of the laser microscopes LMA, LMA1, and LMA10).[131] The objective has two reflecting surfaces, the concave mirror, **(1),** and the coated convex surface of the second component of the lens-mirror combination, **(2).** The objective is protected by a plate, **(3).**

**Table 4.6**  Specifications of Objectives and Lenses of Various Laser Ablation
Atomic Emission Analyzers

| Type | Lateral magnification and numerical aperture | Working distance (focal length) (mm) | References |
|---|---|---|---|
| Flat-field achromat | ×16/0.20 | 18 | Zeiss Jena, Germany[12,130,131] |
| Reflecting flat-field apochromat[a] | ×40/0.50 | 15.7 | Zeiss[12,130,131] |
| Achromat | ×10 | 13 | Jarrell Ash, USA[131 (p. 47)] |
| Achromat | ×10 | (32) | Jarrell Ash[131 (p. 46)] |
| Flat-field achromat | ×5/0.10 | 24.9 | Lomo Leningrad, Russia,[131 (p. 47)] |
| Flat-field achromat | ×8/0.17 | 18.8 | Lomo[131 (p. 47)] |
| Achromat | ×14.4/0.30 | 5.7 | Lomo[131 (p. 47)] |
| Plan-convex lens | | (250 | André et al.[102] |

[a]  See Figure 4.12

analysis in metallic and mineral specimens, since their analytical lines (and also
those of other elements, such as Cl, Br, I, O, H, and N) lie in the spectral region
<200 nm; and (2) lowering the limit of detection in the entire spectral range by
using gas purgings during the ablation, evaporation, and excitation process.
Some of the first results with a sample cell were published by Piepmeier and
Osten in 1971.[148] Some results by other authors and from our own investigations
are reported in reference 117. More recently, Ownes and Majidi[149] described the
optimization of a buffer gas and its pressure to improve the signal-to-back-

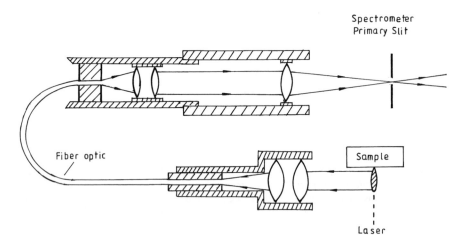

**Figure 4.13**  Optical coupling.[146] Reprinted with permission of the author.

ground noise ratio on LIBS. Sabsabi et al.[103] focused the laser radiation of a pulsed KrF excimer laser (248 nm) by a simple lens of 1 m through a quartz window of a vacuum-tight chamber (1 m³) onto the surface of an Al-Mg alloy at an incidence of 45°. Lee and Sneddon[122] focused the radiation of an ArF excimer laser onto metallic sample surfaces in a laboratory-constructed stainless steel ablation chamber and found that gas breakdown in an excimer laser–ablated plume occurs to a greater extent in an argon atmosphere than in air or helium. Wisbrun[82] studied the effect of gases above the environmental samples. He found that argon, an inexpensive gas, can be used in practical applications in order to increase the signals by a factor of 1.8.

### 4.4.2.4 Dispersing Systems and Detectors

In the 1960's and 1970's photographic recording proved satisfactory for laser ablation. The position and intensity of spectral lines was recorded on long-lasting photographic material. This high detection capacity atomic emission spectrography[117,131] has the ability to store both spectral line position and intensity over a wide spectral range, which permits the bulk of the elements of the periodic table to be recorded simultaneously. It is simple to operate and has a relatively low purchase price and maintenance cost. Its disadvantages are that a chemical process is necessary for the development of the photographic layer (which is time consuming, and may cause problems with reproducibility) and requires qualified personnel and a microdensitometer with suitable accessories for evaluation.

The classical spectrometer principle consists of arranging slits in the focal plane of a spectral apparatus that screens off the lines. Photo multiplier tubes (PMTs) are placed behind the slits as detectors, which produce a photo current proportional to the intensity of the incident radiation. Detection is quick and accurate, but disadvantages remain. Because the device is limited to fixed exit slits and photo multiplier measuring channels, the elements that could be analyzed were either predetermined or the channels had to be adjusted to allow analysis for the desired elements. Both situations require that the composition of the sample be known, which is frequently not the case. This photoelectric recording procedure could not be used universally.

In 1978 and 1979,[150,151] two series of investigations were carried out with combination of a laser microanalyzer and two generations of optical multichannel analyzers, OMA1 and OMA2 from PARC. As connecting units the authors used polychromators with a focal distance of 0.3 m and exchangeable gratings with 600, 1200, and 2400 lines mm (from Jarrell-Ash and McPherson). Depending on the choice of grating, it was possible to record either a large spectral range with low resolution, a medium range with medium resolution, or a small range with good resolution. The OMA1 coupling was made through a 0.3-m Ebert polychromator from Jarrell-Ash. The polychromator was equipped with the user's choice of a 590- or a 2360-line/mm grating. Since the target was 12.5 mm long, the spectral range that could be covered was 70 or 17 nm, respectively. Wisbrun et al. (1994)[82] recorded and resolved their spectra by a gated OMAIII system. Four spectral regions of about 50 nm were used, starting at 217.6, 320.6, 389.5, and 475.1 nm.

During the eighties, advances in charge-coupled device (CCD) technology resulted in linear arrays of greater than 2048 elements, and two-dimensional arrays with up to 380 × 488 elements. The dynamic range was slightly less than that obtained with diode arrays, and because the light was detected through a transparent electrode, the spectral range was more limited.

André et al. (1994)[102] used an f/6 spectrometer equipped with a gated intensified photodiode array (Reticon 1024 photodiodes, 700 intensified photodiodes). The wavelength range simultaneously recorded was approximately 13 nm (at 300 nm) and the spectral resolution, measured over 5 diodes, was 0.1 nm (at 300 nm). Russo et al.[105] reported in 1995 that a charge-coupled device detector should be three orders of magnitude more sensitive than a photodiode array detector.

### 4.4.3 Qualitative and Quantitative Analyses

Laser microanalysis (LMA) was originally developed to enable qualitative local analysis with spatial resolution of material of varying composition, including conducting and nonconducting materials. Since the method was shown to be suitable to this type of analysis, efforts have been made to use LMA in quantitative analysis as well, inevitably requiring a systematic study of the problems.

The necessity of development and optimization of lasers and laser parameters for quantitative analysis is related to the nature of the analytical sample. It is well known that quantitative analysis involves relating the signal intensity of an element in the plasma to the concentration of that element in the target. Whether or not this is true depends on the combined effects of the physical and chemical properties of the target on the plasma composition.

Since laser ablation atomic emission spectrometry is not an absolute method, the problem of preparation, evaluation and testing of standards (i.e., reference materials) arises. One of the main problems is to get homogeneous reference samples or to find other means for standardization. Methods of preparation of homogeneous reference samples and homogeneity tests of such samples are described in reference 117, pp. 76 to 84.

The results of André et al. (1994)[102] demonstrate that LA-AES with internal standardization, when performed in well-controlled experimental conditions, is a quantitative method with a 2% reproducibility for metal and alloy matrices. Russo et al. (1995)[105] analyzed glass samples and found that the precision between five spots after 60 s of pulsing improved to 5–10% relative standard deviation.

### 4.4.4 Application

The fields of application from 1962 to 1989 were reviewed by Moenke-Blankenburg in 1989 (see reference 117, pp. 120–180). Hence, only a short overview, reduced to the period of 1990 to 1995 is given in Tables 4.7 and 4.8.

**Table 4.7** Application of LA-AES/LIBS in Chemical Analyses of Minerals, Rocks, Soils, Sands, Sewage Sludges, Glasses, and Salts

| Elements | Matrices | Object of analyses | Ranges of concentration | LOD (ppm) | RSD (%) | References |
|---|---|---|---|---|---|---|
| Ba, Fe, Mn | Ferromanganese concretions | Distribution analysis | Minor, major elements | | ~10 | Moenke-Blankenburg, Günther[45] |
| Si, Al, Ca, Mg, Al, Fe | Rocks, raw material | On-line control of mass streams | Major elements | | | Lorenzen et al.[147] |
| Cd, Cr, Cu, Ni, Pb, Zn | CRM144, CRM146, PACS-1, soils, sands, sewage sludges | Determination of heavy metals in environmental samples | Minor, trace elements | 10–30 | ~15–80 | Wisbrun et al.[82] |
| Cr, Pb, Be | Soils | Direct determination by two methods | Minor, trace elements | 2 for Cr, Pb 0.1 for Be | 20 | Han, Cremers[152] |
| K | Glasses | Rapid, direct method | Major, minor concentration | | <10 | Lee, Sneddon[153] |
| Cl, Br, I | Salts | Spectral identification | Major elements | $10^{-7}$–$10^{-8}$ g | | Idzikowski, Gadek[154] |

**Table 4.8** Application of LA-AES/LIBS in Chemical Analyses of Metals and Alloys

| Elements | Matrices | Object of analysis | Ranges of concentration | LOD (ppm) | RSD (%) | References |
|---|---|---|---|---|---|---|
| Al, Cu Fe, Pb, Zn | Pure metals | Ambient Ar-gas breakdown in an excimer laser-ablated plasma | Traces | | | Lee, Sneddon[122] |
| C | Liquid steel | Industrial process and quality control | Minor element | 10–100 | 1–2 | Lorenzen et al.[147] |
| C | Liquid steel | Review | Minor element | | | Moenke-Blankenburg[155] |
| Al, Mg | Alloys | LIBS, fundamentals | | | | Sabsabi et al.[103] |
| Al, Mg | Alloys | UV-LA in OES, study | Major elements | | ~2 | André et al.[102] |

### 4.4.5 Principle of LA-ICP-AES

Laser ablation offers a significant potential for the direct introduction of solids into an ICP. Combining the sampling capabilities of laser ablation with the analytical performance of the ICP presents some interesting possibilities. The inductively coupled plasma is one state-of-the-art emission spectrometric excitation source. The high temperature (5000–8000 K), high stability, and relative freedom from matrix effects of this source yield low detection limits for most elements, calibration curves that are linear over four to five orders of magnitude, and a precision of a few percent or less. Because the ICP-AES operates at atmospheric pressure, the introduction of ablated material is easy. The first attempts to combine laser ablation with inductively coupled plasma excitation were done by Abercrombie in 1977,[156] Salin et al. in 1979[157,] Carr and Horlick in 1980,[158] and Thompson et al. in 1981.[29] The principle of LA-ICP-AES is shown in Figure 4.14.

### 4.4.6 Ablation Cells and Transport Processes into the ICP

#### 4.4.6.1 Ablation Cells

The design of the ablation cell is of decisive importance for the ablation event *and* the transport process of the sample material into the ICP.

Thompson et al.[29] published in 1981 the first design of an ablation cell for application in LA-ICP-AES. The cell was made from glass, covered with a plate of optical silica glass 35 mm in diameter with a thickness of 2 mm. The laser beam–transmittable window on the top of the cell was placed above the surface of the sample a distance of less than 17 mm, which was the focal length of one of the objectives of the Laser Micro Analyzer LMA10. In addition, cells of various diameters, from 20 to 50 mm were tested to accommodate different sizes of samples.

Carr and Horlick (1982)[30] kept the connection of the sampling chamber to the torch of the ICP as short as possible. Thus, the sampling chambers utilized in their work were designed to fit inside the shielded plasma chamber directly

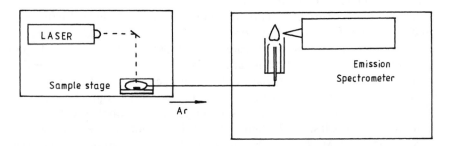

**Figure 4.14** Principle of laser ablation inductively coupled plasma atomic emission spectrometry.

below the torch. Two chambers designed to accept samples of different geometry were built. The first chamber was built to handle samples in the form of rods or tubes $\frac{3}{8}$ inch or less in diameter and from $\frac{1}{2}$ inch to 4 inches in length. To present a new surface for sampling, the sample was moved up and down with a threaded rod screwed into the Teflon base. The upper part of the chamber was constructed out of a machinable glass rod $1\frac{1}{2}$ inches in diameter. The transparent chamber simplified alignment, sample rotation, and focusing of the laser. The laser beam was focused on the rod surface with a short focus of a 5-cm spherical lens. A disc sampler was designed to accommodate discs from $\frac{1}{2}$ inch to approximately 5 inches in diameter and from $\frac{1}{32}$ to $2\frac{1}{2}$ inches in thickness. As with the rod sampler, the disc sampler mounts inside the plasma chamber directly beneath the torch, and the laser beam is focused through the optical window onto the sample surface.

Ishizuka and Uwamino (1983)[31] used an ablation cell made of a stainless steel wall with an inner diameter of 13 nm and a height of 35 mm, and a Pyrex glass window. An ablation cell with larger volume (inner diameter 40 mm, height 30 mm) was also examined.

Su and Lin (1988)[159] compared two types of sample chambers. Both the cells were made of soda glass, but with different geometric configurations. The argon inlet and outlet of type I are on the same horizontal line, whereas in type II, the argon is introduced tangentially into the chamber, below the output. For type I with a quadratic cross section of $25 \times 25$ mm$^2$ and a length of 27 mm, the laser glass window is fixed at the argon inlet tube (diameter of 5 mm), but for type II, the glass window (diameter 2 mm) is sealed beside the argon inlet (diameter 10 mm). Type II consists of two separable parts: the lower cylinder (diameter 40 mm) is fixed and sealed to a panel of the gear system. The sample pellet was placed on the top of a graphite platform, which was mounted on the gear system. The sample was shot by the laser beam in a horizontal direction, the beam being focused on the upper rim of the pellet with a 5-cm focal lens. On activation of the laser, rotation of the platform with the sample pellet was commenced and it was moved upward automatically in order to present new areas for continuous sampling. The argon outlet is arranged at the top of the bottle-shaped cell. This cell, type II, improved the detection power by more than a factor of ten compared with type I.

Moenke-Blankenburg et al. (1989)[39] and Günther and Gäckle (1990)[42] proposed two types of cylindrical ablation chambers for LA-ICP-AES, where the sample could be placed inside (Figure 4.15) or outside the chamber (Figure 4.16). The volume of the first cell is variable from 0.25 to 64 cm$^3$ by the insertion of three different rings. The second cell has an open end and fits tight to the smoothed sample surface. Nozzles were used to control the argon flow velocity.

Russo et al. (1995)[105] developed a sampling chamber coupled with a nanosecond KrF excimer laser at 248 nm with a 10-Hz repetition rate and also with a picosecond Nd:YAG laser at 266 nm and 10 Hz. A diagram of the experimental setup and a cross-section scheme of the sampling chamber is shown in Figures 4.17 and 4.18. The laser beam is focused onto the analytical samples

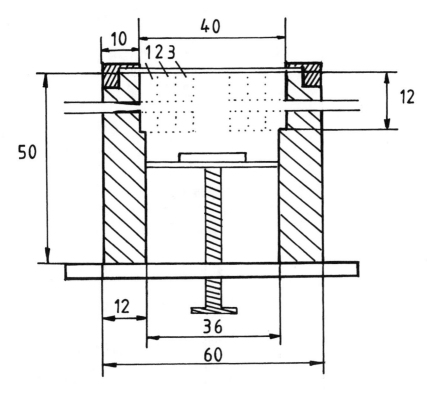

**Figure 4.15** Sample chamber (1989).[39,42]

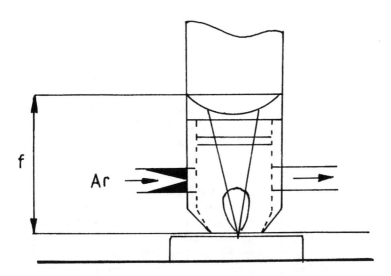

**Figure 4.16** Sample chamber (1990).[42]

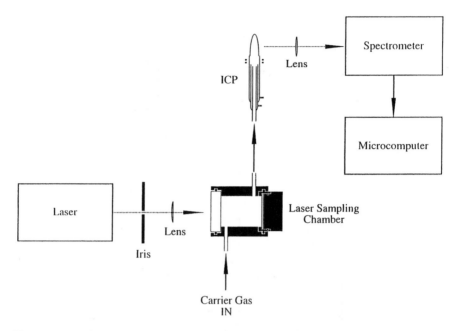

**Figure 4.17** Diagram of the laser ablation ICP-AES experimental configuration. The chamber is mounted on an *x-y-z* translation stage for focus and lateral adjustments. A nanosecond KrF excimer laser (248 nm, 30 mJ) and picosecond Nd:YAG laser (266 nm, 10 mJ) were used.[105] Reproduced with permission of the Royal Society of Chemistry, Cambridge.

with a plan-convex lens with a 20-cm focal length. The effective focal length is 18.3 cm due to the higher refractive index of the lens (fused silica) at the UV wavelengths. The chamber has an outer diameter of 3.18 cm, an inner diameter of 2.54 cm, and an internal length of 3 cm. The window is detachable for replacement, because the high laser fluence and deposits of sample material can damage or contaminate it. Window deposits are minimal because the carrier gas flows from the window toward the sample. The sample holder is water cooled. The cell, mounted on an *xyz*-translator, is sealed from the atmosphere with O-rings at the window and at the sample holder. The substance to be sampled is introduced into the ICP through a Teflon tube 20 cm in length and 3 mm in diameter.

It should be of interest to see what is new in the G. Horlick's group since their development in 1982[30] of the shortest connection between laser, sample, and ICP. Liu and Horlick (1995)[160] have proposed a system that eliminates the transport step. The sample is placed inside the torch about 1 mm below the plasma discharge (Figures 4.19 and 4.20). The focused laser beam is directed through the plasma from above the discharge to the sample, which is positioned in a direct sample insertion device (DSID). This system was described in detail by Liu and Horlick in 1994.[161] Now, the results of the LA-DSI-ICP-AES system

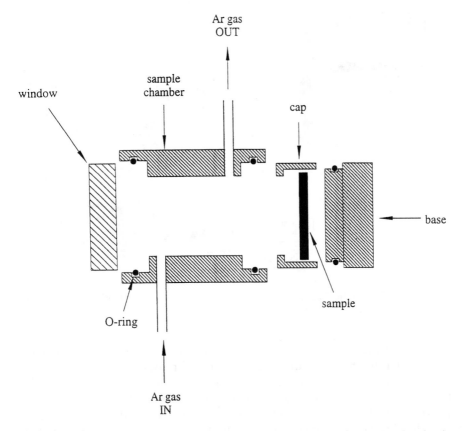

**Figure 4.18** Detailed diagram of the laser ablation sampling chamber. The chamber is a machined-out brass rod. The window is a 1-mm thick quartz plate.[105] Reproduced with permission of the Royal Society of Chemistry, Cambridge.

were compared with the results of a conventional LA-ICP-AES system (Figure 4.21a) with a transportation tube of 1 m and a sampling cell (Figure 4.21b).

### 4.4.6.2 Transport Processes

The ablated material produced from the sample surface by laser shot is mixed in the ablation cell with the carrier gas (e.g., argon), then entrained in a tube and transported into the inductively coupled plasma, the secondary source for subsequent excitation and/or ionization.

Since ablation, transport, and secondary excitation/ionization take place sequentially, each may be independently optimized; secondary excitation of atoms as well as ion formation are more or less insensitive to the physical and chemical state of the ablated material.

In LA-ICP-AES the height of the signals depends on the length of the tubing,

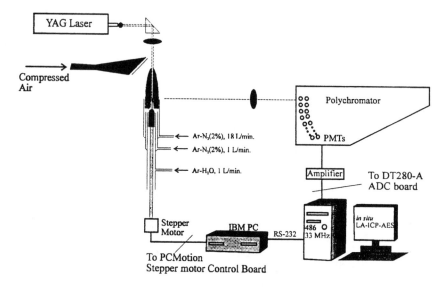

**Figure 4.19**  Schematic diagram of the in situ laser ablation inductively coupled plasma atomic emission system.[160] Reprinted with permission of Elsevier Science, Amsterdam.

as shown in Figure 4.22,[39] where the transport distance between the chamber and the ICP torch was modified from 0.5 to 2.5 m (the inner diameter of the PVC tube was 5 mm; the argon transport gas flew through the chamber with a flow rate of 0.7 L/min). The increase in length also resulted in peak broadening. Ishizuka and Uwamino[31] compared the effect of the length of the tubing on peak height and peak area values. Both intensities decreased gradually with increasing length. With the laser in each mode, the degree of the decrease in intensity with increasing length was larger for the peak height value than the peak area value. They assumed that this must be due to the diffusion of laser-ablated sample material in the PVC tubing. An attempt was made to compare the aerosol transport with the transport mechanism of solutions through flow injection analysis.[39]

The following can be concluded:

- The dispersion decreases with decreasing tube length
- The dispersion decreases with decreasing tube diameter
- The dispersion decreases with the use of packed beads in the tube[39]
- The dispersion decreases with increasing aerosol density (sample volume)

There is now abundant evidence that material mobilized by laser ablation has most of its mass in particles that range in size from 0.5 to 5 μm.[30,34,35,40,162-165] The average of transport efficiency is about 40%.[119] In an ideal system, it should be expected that the elemental composition of the aerosol after transport and thermal treatment in the secondary source would mirror the composition of the sample. Recent studies have shown that there are problems even with apparently

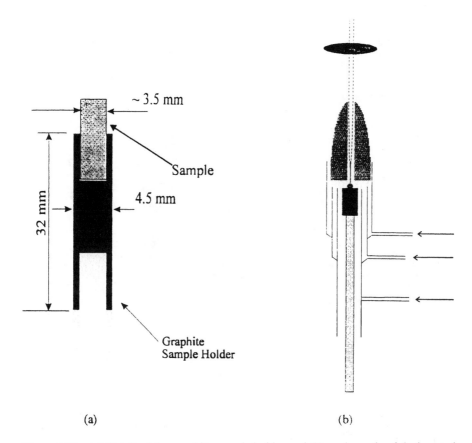

**Figure 4.20**  **(a)** Detail of the graphite sample holder and **(b)** a schematic of the inserted sample probe.[160] Reprinted with permission of Elsevier Science, Amsterdam.

simple materials.[40,165] In the first of these studies, pyrite and stainless steel were ablated and the ejected material determined by energy-dispersive X-ray spectrometry. For pyrite, it was found that S was lost. For steel, it was found that the composition depended on the mode of laser action. For the Q-switched mode and, with some exceptions, the free-running mode, the particles had the same composition as the bulk material. The exceptions were thought to arise from melting of inclusions. The second study[165] concerned two minerals, pyrite and olivine and used three laser systems: ruby at 694 and 347 nm, and Nd:YAG at 532 nm. In summary, the results revealed the production of particles with different chemical compositions, which is an indicator of considerable complexity in heating regimes, melting, vaporization, and thermo-mechanical shock.

### 4.4.7  Instrumentation of LA-ICP-AES

The principal technical elements required for LA-ICP-AES are a suitable laser, suitable focusing optic, suitable ablation chamber, interface connections that

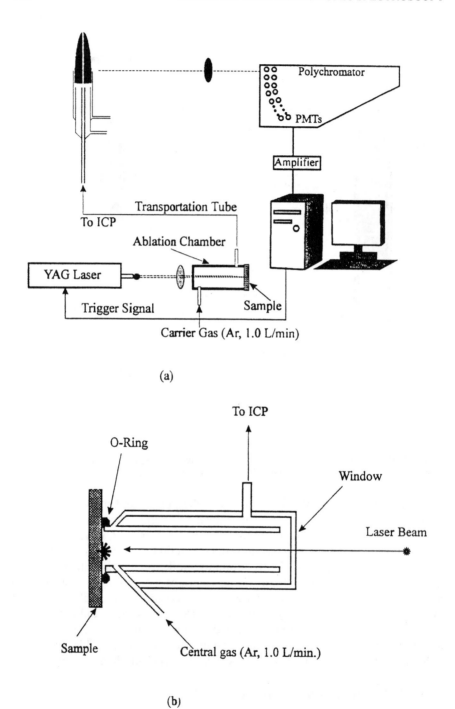

(a)

(b)

**Figure 4.21** Schematic diagram of the conventional laser ablation setup, **(a),** and detail of the ablation cell, **(b)**.[160] Reprinted with permission of Elsevier Science, Amsterdam.

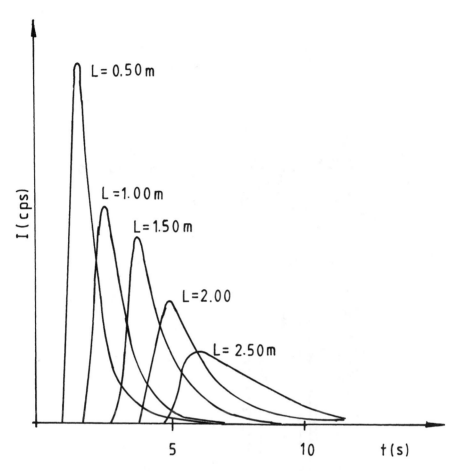

**Figure 4.22**   Variation of transport distances from 0.5 to 2.5 m between the sample cell and the ICP torch (in LA-ICP-AES). The inner diameter of the PVC tubing was 5 mm.[39,42]

ensure efficient transport of the aerosol into the plasma, and a simultaneous multichannel emission spectrometer.

The insertion of the tube-transported laser-ablated and predominant vaporized material into the ICP torch is in most cases done through the normal aerosol channel, for instance, using a three-way tap below the torch (see, for example, references 29 and 39).

A typical argon ICP torch used in the current ICP instruments consists of an assembly of three concentric quartz tubes. Figure 4.23 shows the present-day form of a Fassel torch,[166–168] which features two tangential gas entries to the outer and intermediate tubes. Aerosol is introduced through the inner injector tube along the torch axis. Typically, the injector tube has an orifice of 1.2–3 mm. Argon gas is used almost exclusively in the gas streams, although other

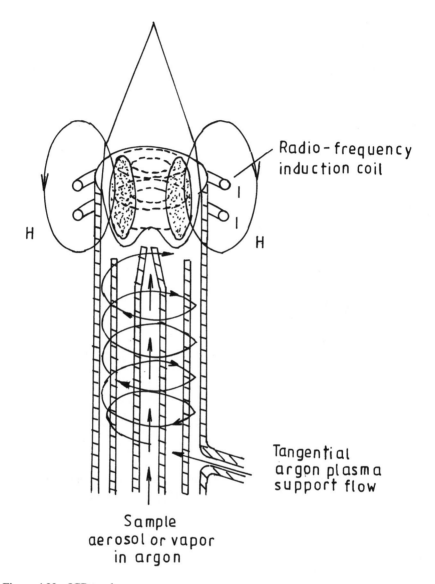

**Figure 4.23**  ICP torch.

gases and gas mixtures are increasingly used in the ICP to raise plasma temperatures and improve detection limits.

### 4.4.7.1 Spectrometer

In atomic emission spectrometry several different kinds of spectrometer are used. The Paschen-Runge mounting[169] is presently the most commonly used optical configuration for direct-reading spectrometers. In this mount, the grating and slits are in fixed positions, with exit slits mounted along the Rowland circle. In an Ebert-Fastie polychromator,[170,171] the entrance and exit slits are on either side of the grating, and a single concave spherical mirror is used as a collimating and focusing element. The Czerny-Turner variant of the Ebert mounting incorporates two smaller concave mirrors instead of a single mirror. The simplest scanning mechanism is achieved by rotating a grating about a vertical axis through the center of the grating. In the Seya-Namioka mounting,[172,173] this is accomplished by using a concave grating and keeping the angle between the incident and diffracted beams constant angle at about 70°. Figure 4.24[174] shows the layout of the optical system of an echelle grating spectrometer, an instrument highly efficient in separating the complex, line-rich spectra from an ICP source. Light from the entrance slit is directed by the collimating mirror to the echelle grating, and the diffracted beam then is reflected to the cross disperser system with a prism and lens that acts as an order-sorter and focusing lens, forming a two-dimensional spectral pattern in the focal plane of the spectrometer. The echelle polychromator for ICP-AES was designed for use with the multichannel solid-state detector described by Barnard et al.[174] The optical system has separate output sections for the UV and visible wavelength ranges, which permit high spectral resolution and high optical throughput while cov-

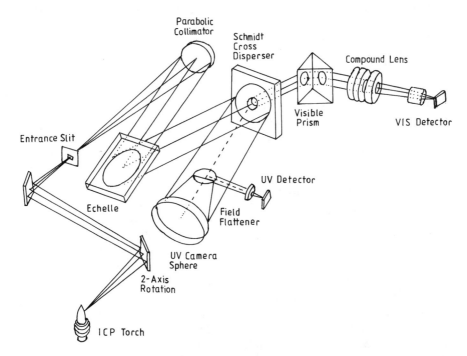

**Figure 4.24** Layout of an echelle-grating optical system for ICP-OES.[174] Reproduced with permission of the American Chemical Society.

ering the spectral range from 167 to 782 nm (applications of this system are described in references 46, 79, and 80). A recent alternative approach has been to transfer light from the focal plane to detectors by optical fibers.[175] This spectrometer (see reference 47 for applications) consists of a high-aperture, high-resolution, 0.5 m-echelle monochromator and 132 light guides arranged behind an exit mask to transfer light to 12 photomultiplier tubes.

### 4.4.7.2 Detection Systems

The main detectors used in atomic emission spectrometry today are photomultiplier tubes (PMTs), photodiode arrays (PDAs), charged-coupled devices (CCDs), and vidicons, image dissectors, and charge-injection detectors (CIDs). The advantages and disadvantages of PMTs were mentioned above under the topic Instrumentation of LA-AES. Self-scanning linear silicon photodiode arrays have been used successfully in atomic emission spectrometry since 1976.[176-179] Recent development of PDAs for ICP-AES has been reviewed by Lepla and Horlick.[177,178] Arrays with 1024, 2048, and 4096 elements are commercially available. Currently, the 1024-element configuration is the most commonly used. The main advantage of a PDA is the fact that it measures spectra simultaneously, but the wavelength coverage is quite limited, for example, for moderate- and high-resolution studies it is about 10 nm unless a special spectrally segmented photodiode array spectrometer is used.[79,80] In this case, the array detector allows the analyst to gain, simultaneously, a wealth of spectral information, comparable to that from a photo plate, with the ease of operation associated with photomultiplier tubes.[180] Array spectrometers are especially suitable for tandem techniques, because transient signals (Figure 4.25) are being recorded. The ability of the array detector to measure background correction points simultaneously,[181] and possibly internal standards, is of vital importance. Since the matrix composition varies, the risk of spectral interferences cannot be ignored. Therefore, the multiline approach of the array spectrometer adds stability to the method, along with the good resolving power of the echelle-based optic.

### 4.4.8 Methods of Quantitation

Calibration in LA-ICP-AES is usually carried out by using solid external standards, commercially available or self-made, or by using the standard addition method, if an addition is possible. There are three principle problems that must be taken into account: fluctuations, especially of the ablation and excitation process; matrix effects, depending on the sample composition; and chemical inhomogeneity of the standard materials.

The influence of fluctuations of operating parameters with time on the signal of the analyte can be reduced by applying the ratio of two elements: the analyte and the reference element, assuming each signal is affected to the same extent and the ratio remains constant. For an internal standard, an element of known and steady concentration can be used. Occasionally in compact solids a main

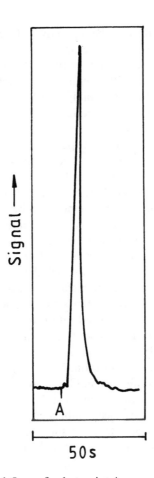

**Figure 4.25** Transient signal. Laser fired at point A.

element is suitable. In pressed powder pellets or fused samples solidified as glass beads, a constant quantity of a pure substance can be added to the samples.

Internal standardization has been used to compensate for the pulse-to-pulse variations in the amount of ablated material.[182–187] Richner et al.[188] used an optical method to correct for such variations. When applied to LA-ICP-AES, external standards, which are commercially available, suffer from physical and chemical matrix effects. A physical matrix effect refers to the influence exerted on the removal process by the nature of the sample in its widest physical sense. Chemical matrix effects originate in the dependence on the chemical composition of the matrix and its influence of the ablation, transport, and excitation conditions of the analyte.

Furthermore, external standards suffer from inhomogeneity in microregions. Chemical homogeneity is a relative property of a solid which depends on the spatial resolution and precision of the analytical procedure being used for in-

vestigation. Solids are defined as being analytically homogeneous if the variations in the chemical composition over the entire sample volume determined in
various areas of the sample are not significantly larger than the error of the
analytical procedure. Moenke-Blankenburg and Pacher[189] tested different samples with laser microanalysis for suitability in laser microanalysis by testing an
area and a profile by statistical methods. According to Danzer and Küchler[190]
the aim of homogeneity tests should be to find the limit of homogeneity for
each element in a standard for a given spatial resolution and a known precision
of the analytical procedure. The standard addition method is useful for LA-
ICP-AES of pressed pellets or glass beads by fusing the sample (i.e., for distribution and bulk analysis only). This method, where the powdered analyte is
spiked with known amounts of the elements to be determined, can be utilized
to minimize the errors of matrix effects and inhomogeneity. Both classical methods are limited for application in laser microanalysis. A new approach for quantitation was proposed by Baldwin et al. in 1994.[78] During the laser ablation
sampling process, solution standards are nebulized and the aerosol is added to
the laser-ablated aerosol to generate a standard addition curve for the analyte
being determined. The method combines the solution standard addition with
aerosol mass measurement for normalization of the process. The diagrams of
the instrumental setup and the aerosol flow path are given in Figures 4.26 and
4.27. Thompson et al. 1989[191] carried out a study to assess the possibility of
using aqueous standards for calibration. Moenke-Blankenburg presented a paper at the Third Surrey Conference in Guildford, UK, in July 1989[39a] and proposed the liquid-solid calibration technique. Since then, some successful applications of this technique have been published.[44-47] The principle is described
here. In the first step, aqueous standard solutions are nebulized in the normal
manner, but carried by only one part of the divided Ar stream. The other part

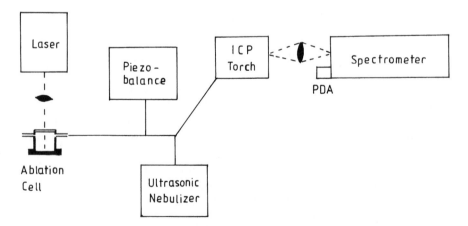

**Figure 4.26**  Block diagram of the instrumental setup used for the aerosol mass measurement and solution standard addition LA-ICP-AES technique.[78] Reproduced with the
permission of the American Chemical Society.

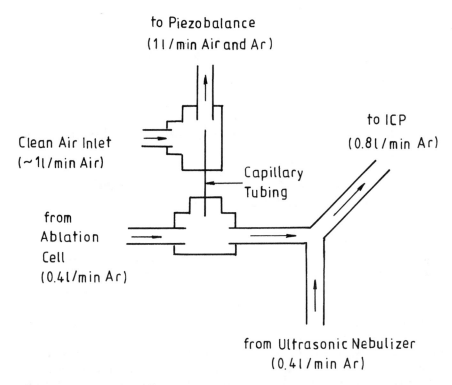

to Piezobalance
(1 l/min Air and Ar)

Clean Air Inlet
(~1 l/min Air)

to ICP
(0.8 l/min Ar)

Capillary
Tubing

from
Ablation
Cell
(0.4 l/min Ar)

from Ultrasonic Nebulizer
(0.4 l/min Ar)

**Figure 4.27**   Schematic diagram of the aerosol flow paths. Approximately 5% of the argon and laser-ablated sample aerosol are drawn into the piezobalance through the capillary tubing. Makeup clean air is drawn into the piezobalance to maintain a flow rate of 1 L/min. The remainder of the laser-ablated aerosol is added to the ultrasonic nebulizer aerosol and introduced into the ICP torch.[78] Reproduced with permission of the American Chemical Society.

of the gas stream is allowed to flow through the laser ablation chamber, but without laser action. In the second step, intensities for the laser-ablated solid (analyte and reference element) are measured using both flow streams: one flowing through the ablation chamber transporting the aerosol, and the other transporting a blank solution. The two Ar streams are mixed in front of the ICP torch so that the water or liquid introduction into the plasma is the same, thus maintaining similar plasma conditions. The desired analyte concentration is calculated by the formula

$$c_{sa} = (c_{la}/c_{lr})c_{sr} \tag{3}$$

where, $c$ is the concentration, sa refers to the solid analyte, la and lr stand for the liquid calibration samples of the analytical element and the reference element, respectively, and sr is the solid reference element.

### 4.4.9 Application

The fields of application up to 1993 have been reviewed by Moenke-Blankenburg[28] and Darke and Tyson in 1993/94.[118,119] Examples for application of LA-ICP-AES after 1993 are summarized in the Tables 4.9 and 4.10.

## 4.5 Laser Ablation Atomic Mass Spectrometry and Laser Ablation Inductively Coupled Plasma Mass Spectrometry

### 4.5.1 Principles and Performances of Laser Ablation Atomic Mass Spectrometry

#### 4.5.1.1 LM-MS since 1963

The first use of a solid-state laser for laser micromass spectrography (LM-MS) was reported by Honig and Woolston as early as 1963.[13] Since the demands made on both the performance of the laser and on the focusing optic were similar to those in laser microemission analysis, it was natural to supplement a laser microanalyzer with an ion source and to couple it with a suitable mass spectrograph. The starting point for these experiments was the interest in extending local analysis in the microregion to the determination of isotopes in order, among other things, to open up new possibilities for tracer technique in solid-state research and in isotope analysis in geology and biology. The first laser of the first generation apparatus used, a ruby laser, delivered an output energy of 1 J with a pulse duration of 200 µs, and a power of 5 kW.[13] Therefore, it was suitable for ionizing all elements with an ionization potential of $\leq 10$ eV in the microplume. Trace analysis in the parts per million region and isotope analysis could be performed in microregions of solid surfaces down to $10^{-4}$ cm$^2$. In the first years, spectrographs with photoplate detection were used. Afterward, time-of-flight (TOF) spectrometers were introduced.[14–27] A survey of parameters and performances of LM-MS between 1963 and 1972 is shown by Moenke-Blankenburg in Chapter 9, Table 45 of reference 117. The second generation of laser microprobe mass analyzers was developed from the middle of the 1970s to the beginning of the 1980s. Two major areas of laser mass spectrometry for analytical purposes have been evolved during the last decade: microprobe analysis with a spatial resolution of approximately 0.5 µm and a sensitivity in the parts per million to parts per billion range, and laser desorption of thermally labile, nonvolatile organic compounds.[200, 201] All laser microprobe mass analyzers that were commercially available during this period were based on essentially identical functional principles. They used Nd:YAG lasers of wavelengths of 1064 nm and, later on, wavelengths in the UV, typically the quadrupled wavelength of Nd:YAG at 266 nm. Now, excimer lasers with about 7–30 nm pulses at wavelengths of 351 nm (XeF),[100] 308 nm (XeCl),[202] and 248 nm (KrF)[98,203] are introduced. Standard optical microscopes for sample observation and focusing

**Table 4.9**  Application of LA-ICP-AES in Mineralogy, Geology, and Related Fields

| Elements | Matrices | Object of analysis | Ranges of concentration | LOD (ppm) | RSD (%) | References |
|---|---|---|---|---|---|---|
| La, Ti (int.St.) | Pseudobrookite | Liquid-solid calibration | Trace | | ~4 | Moenke-Blankenburg, Günther[45] |
| U, Si (int.St.) | Soils | In situ field and lab. anal.; comparison | Trace | 0.01 | Confidence intervals given | Zamzow et al.[89] |
| Al, Ca, Cu, Fe, K, Mg, Mn, Si, Ti | Ores of Al and Mg (NIST) | Comparison of LA-ICP-AES and LIBS | Major, minor | | Confidence intervals given | Thiem, Wolf[192] |
| Fe, Mn, Cu, Zn, Na, K, Ca, Sr | Magmatic ore fluids | Metals in fluid inclusions | Major, minor | | Precision on ratios are given | Wilkinson et al.[193] |
| Fe, Mn, Zn, Cu, Pb, Sn, K, Na | Topaz | Metals in fluid inclusions | Major, minor | | | Rankin et al.[194] |
| Fe, Mn, Cu, Pb, Cr, Zn, Ni | Soils | Heavy metals in soils; calibration | Major, minor, trace | | 4.9–12.7 | Moenke-Blankenburg et al.[46] |

(continued)

**Table 4.9** Application of LA-ICP-AES in Mineralogy, Geology, and Related Fields (*Continued*)

| Elements | Matrices | Object of analysis | Ranges of concentration | LOD (ppm) | RSD (%) | References |
|---|---|---|---|---|---|---|
| Al, Fe, Mg, Mn, Sc, Sr, Ti, Zn | Soils | Test of OPTIMA 3000 with Nd:YAG | Major, minor, trace | | 0.3–8.5 | Nölte et al.[79] |
| Ag, As, Cd, Co, Cu, Fe, Mn, Ni, Pb, Sn, Zn | Galena, Sphalerite, Pyrite | Comparison of LA-ICP-AES and spark ablation ICP-AES | Major, minor, trace | | ~9, LA; ~5, spark | Moenke-Blankenburg et al.[47] |
| Al, Ba, Ca, Cr, Fe, K, Li, Mg, Mn, Na, Ni, P, Si, Ti, Zn, Zr | Glasses | Aerosol mass measurement and solution standard addition; quantitation | Major, minor, trace | | ~10 | Baldwin[78] |
| Mn, Fe, Si, Al, Mg | Glass prototypes | Test of modern setups | Major, minor | | 1–3 | Russo et al.[105] |

**Table 4.10** Application of LA-ICP-AES in Metallurgy and Related Fields

| Elements | Matrices | Object of analysis | Ranges of concentration | LOD (ppm) | RSD (%) | References |
|---|---|---|---|---|---|---|
| C | Liquid steel | Review | | | | Moenke-Blankenburg[155] |
| Ni, Cu, Zn, Mn, Fe, Mg, Al | Steel, brass, glass | Peformance of LA-ICP-AES | Major | | 0.3–1.6 | Russo[195] |
| Cu, Zn, Al, Zr, Fe, Ta, Sn | Brass, alloys | Mass removal versus power density | | | | Shannon et al.[196] |
| Cu | Copper | Temperature, spatial profile of LIP | | | | Mao et al.[197] |
| Cu | Copper | Picosecond (PS) laser, He/Ar | | | | Mao et al.[198] |
| Al et al. | Steel (ALCOA SS319-E6) | In situ: DSI versus tubing | Minor, trace | 0.01 (Ca)–45 (Si) | | Liu, Horlick[160] |
| Cu | Copper | IR laser versus UV laser | | | | Geertsen et al.[199] |

of the laser beam with a spatial resolution of about 0.5 μm for the analysis of thin specimens and about 3 μm for surface analysis of bulk specimens are mostly part of the equipment. Typical power densities in the laser focus on the sample surface range from $10^7$ to $10^{11}$ W cm$^{-2}$. This beam is capable of ablating, atomizing, and ionizing solid materials which may then be directed into the mass analyzer and detected. The characteristics of LM-MS, also called LAMMA (laser microprobe mass analysis), had already made it an attractive technique for single-cell elemental analysis in the eighties,[204–206] and this continued into the nineties.[207] The early development of LAMMA 500 was improved for bulk analysis, called LAMMA 1000. It incorporated a sample chamber that allowed analysis of samples measuring up to several centimeters in each dimension. The LAMMA 1000 and another instrument, named LIMA, in its infancy delivered results with relative standard deviations of 20%, at best. The main problem of LA-MS (laser ablation mass spectroscopy) is the quantitation caused by an irreproducible sample ionization process. It has been stated that quantitative analysis is possible only if the laser is stabilized and the sample is homogeneous. Some attempts have been made to improve the quantitation of LA-MS.[208–210] The very valuable review article of Darke and Tyson[118] presents in its Table 5 the instrumental features and results from 112 authors from 1963 to 1991. Another review (1991) and a bibliography (1991) are also available on this subject.[200,201]

### 4.5.1.2 LA-RIMS since 1977

Besides laser ionization mass spectrometry, laser resonance ionization mass spectrometry (RIMS)[211–214] provides possibilities for ultrasensitive mass spectrometry of solids. The resonance ionization process, sometimes referred to as resonantly enhanced multiphoton ionization spectrometry (REMPI),[215] makes use of the resonant absorption of photons by allowed transitions. By the absorption of the first photon, the material is evaporated and brought to an excited state, and absorption of the second photon brings the free atom to the region of the ionization continuum, producing an ion of a given kinetic energy. The selectivity created ion is subsequently analyzed in a mass spectrometer where additional selectivity is added. A time-of-flight mass spectrometer is well suited for use in combination with pulsed laser excitation and has the advantages of yielding all isotopes resulting from one single laser shot as well as a high transmission efficiency.

Figure 4.28 shows a schematic diagram of a measuring setup with a high-resolution reflectron-type time-of-flight mass spectrometer.[216] An excimer laser was incident on the sample in the horizontal plane, and the mass reflectron deflected the ions in the vertical plane. Bakos et al.[216] irradiated the surface of the sample by an incidence angle of 45° with the focused beam of an XeCl excimer laser. The laser intensity was optimized to reach relatively high sensitivity with low initial energy spread of the ions, which enables high mass resolution. The generated ions are accelerated perpendicularly to the surface of the

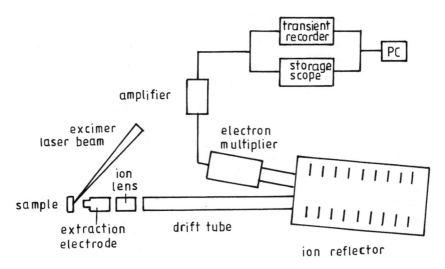

**Figure 4.28**  Schematic diagram of the measuring setup with the reflectron-type time-of-flight mass spectrometer.[216] Reproduced with permission of the Society of Photo-Optical Instrumentation Engineers.

sample by an extracting electrode in the drift tube. An electrostatic lens is used to collimate the ion beam along the flight pass. After the first drift region, the ions enter the electrostatic ion reflector, which consists of rings at equal distances and equally distributed potentials between the potential of the drift tube and the end plate of the reflector. The ions reflected into the second, shorter drift region are detected by an electron multiplier. This setup followed earlier investigations of the same group.[217] The beam of a Q-switched ruby laser was focused onto the surface of the sample. The laser intensity was optimized to $4 \times 10^7$ to $10^8$ W cm$^{-2}$ to atomize the surface effectively and nonselectively and at the same time to reduce the immediate ionization. The evaporated atomic cloud was illuminated by a dye laser beam tuned to the transition of the element to be determined (Na, D$_2$ line at 589 nm, $3^2S_{1/2} - 3^2P_{3/2}$) and an excimer laser beam (which was split out from the pumping beam of the dye laser) at the same instant and spatial volume a few centimeters away from the sample surface. These pulses were synchronized to the ruby laser pulse by an adjustable delay, and the delay time was optimized to illuminate the atomic plume at the peak of its velocity distribution. The generated ions were accelerated by a pair of electrodes to an electron multiplier, and its signal was digitized by a transient recorder.

All results published since (see reference 215) demonstrate the high sensitivity as well as the extraordinary selectivity of LA-RIMS. Nevertheless, separation of the laser sampling from spectroscopic detection does provide new opportunities for optimization and improved analytical performance as described in the following chapter.

### 4.5.2 Principle of LA-ICP-MS

The past decade has introduced laser ablation as a new and versatile solid sampling technique for ICP mass spectrometry. The principles of a laser ablation setup and an ICP mass spectrometer, the most commonly used, are shown in Figures 4.29 and 4.30. As in LA-ICP-AES, samples are mounted on a small turntable or motorized *x-y-z* stage in a glass or steel housing with their surface at the focus of a laser beam. The focused laser radiation causes material to be ablated, sputtered, and vaporized from the surface. The plasma injector gas flow is passed over the sample surface and transports the aerosol through a plastic tube to the ICP torch, where the atomization-excitation-ionization processes ensue.

The potential of a laser system for ICP-MS analysis was first reported by Gray[218] in a paper describing the use of a ruby laser. Arrowsmith[219] was the first to report the use of an Nd:YAG laser with an ICP-MS. In these early applications a single laser shot of about 0.5 J on the sample produced a pit about

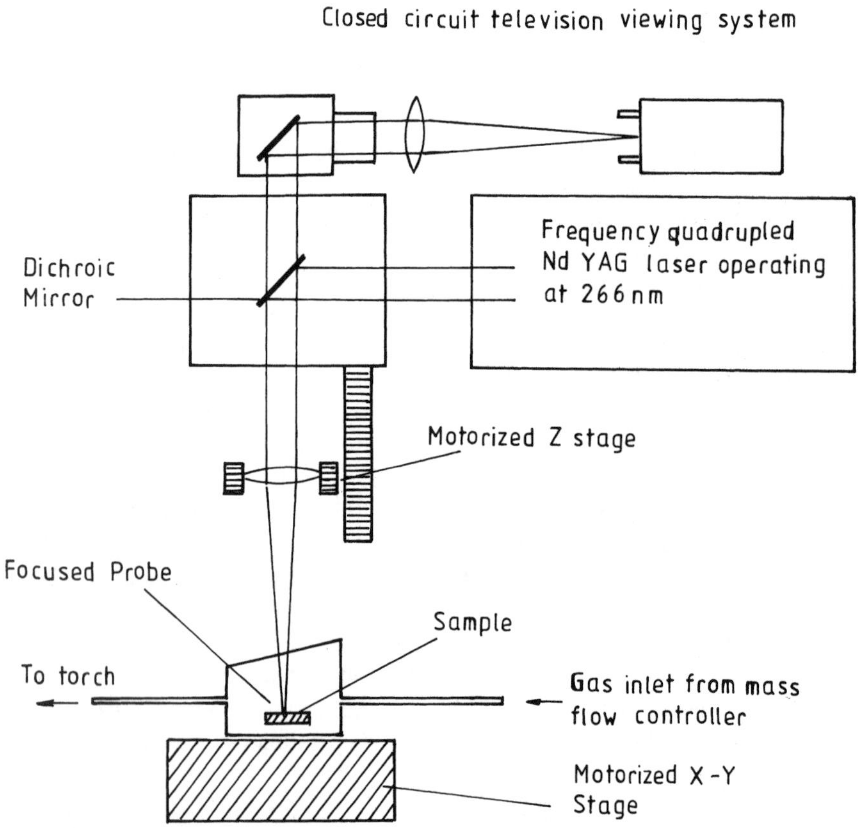

**Figure 4.29**   Schematic diagram of a laser ablation part of LA-ICP-MS.

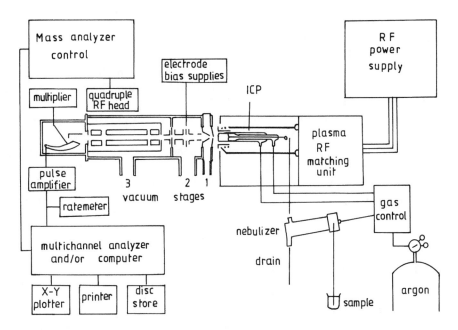

**Figure 4.30**   Schematic diagram of an ICP mass spectrometer, quadrupole type.

0.5 mm in diameter and about 0.5 mm deep, removing about 0.2 mg of material, suitable for use in bulk sampling applications. If the mass analyzer was set to a single ion mass, each laser pulse produced a transient response. For multi-element analysis or isotope ratio determination, the duration of the response peak was sufficient for the completion of many scans at the usual rate of $10 \text{ s}^{-1}$. The availability of integrated LA-ICP-MS systems from several manufacturers has done much to stimulate research and further promote the technique.

### 4.5.3  Instrumentation of LA-ICP-MS and Operating Parameters

#### 4.5.3.1  Laser, Microscope, and Optical Arrangements

Laser types suitable for ablation of solid samples are described in the first part of this chapter. There are no special needs for coupling all the laser types with ICP-MS instead of ICP-AES. Laser microscopes and objectives were developed in the early sixties,[7,12] particularly for the 694-nm wavelength of the ruby laser. The objective, shown in Figure 4.12, is a two-mirror system, which consists of a concave and a coated convex mirror as the second component of the lens-mirror combination. The objective was calculated for the eye-sensitive wavelength of 550 nm, which is not far from 694 nm. The glass parts of the objective were made from optical glass that absorbs ultraviolet wavelengths. The use of UV-laser radiation requires the modification of the optical components from glass to quartz and adaptation of the reflectivity of the coatings to the short wavelengths. Such a modified objective was used by Cheatham and White[84] in 1994. A special 25× reflective objective of an Olympus petrographic microscope was capable of handling the broad spectrum of laser wavelengths from 1064 nm to 266 nm. This objective has a working distance of 14 mm. The microscope is outfitted for both transmitted and reflected light viewing. Chenery and Cook[92] used a high-quality Leitz microscope linked to a Nd:YAG laser operating in the UV at 266 nm. The LAM system of Longerich et al.,[57,220,221] is shown schematically in Figure 4.31. It consists of a Q-switched 1064 nm Nd:YAG laser, a petrographic microscope, a sample cell, and a laser power meter. Brenner and Zhu reported on a special Geolaser probe with promising prospects.[222] CETAC announced a microscope with reflected and transmitted lighting, automatic zoom optics, and a video camera.[223] An auto-focus system for reproducible focusing of laser beams is recommended by Cousin et al.[224]

### *4.5.3.2 Chambers and Transport Processes*

Several cell designs have been systematically evaluated for gas flow entrance and transport of laser ablated material in connection with the development of LA-ICP-AES. In the most common design, the sample is enclosed by the cell.[29,31,39,42,50,159,218,219] Desirable features of this type of cell include minimal volume (but variable to accommodate a range of different sample types and geometries), capabilities for translation and rotation of samples, provision for laser alignment, focusing, and remote viewing of samples, flexibility and convenience for analysis, high transport efficiency, and low memory effect. In the first reported application in ICP-MS, Gray[218] employed a cylindrical glass ablation cell. The sample was placed in a cup on a swing-ridge, argon was introduced tangential to the cell and the sample surface. Figure 4.32[91] shows an open-ended sample chamber constructed by Arrowsmith and Hughes.[225] The sample is enclosed entirely by the cell, but the cell does not form a seal to the sample surface. An outer box is a gas tight and filled with an inert gas to provide a pressure gradient along the transport line. The advantage of this is that the size and shape of the sample is nearly unlimited. As a disadvantage, deposition of sample material on the walls, the window, and the outlet may be expected. Three different ablation cells were tested.[225,226] All have cylindrical geometry and were constructed in glass to provide smooth walls. Glass or quartz windows are attached with silicone sealant for easy replacement and cleaning. The internal volume ranged over 1 to 3 cm³, internal diameters were 5–10 mm, and heights were 20–40 mm. A closed sample chamber was developed by Cousin and Magyar in 1994.[50] The cell has a minimal dead volume. The gas outlet is below the cell window, so less deposition on the cell window results. A laser ablation cell shown in Figure 4.33 was designed and fabricated in-house by Prabhu et al.,[86] originally for studies on laser ablation of solids. The cell was then modified to accept small volumes of solutions. About 200 µl of liquid analytes were placed in a small PTFE boat, and a graphite wheel, partially dipped in the solution,

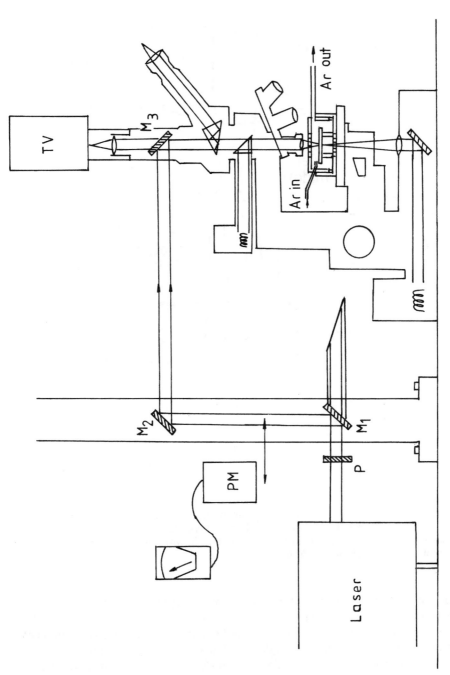

**Figure 4.31** Laser ablation microprobe solid sample introduction system. $M_1$, beam sampler; $M_2$ and $M_3$, mirrors; P, half wave plate; PM, power meter.[220] Reproduced with permission of the authors.

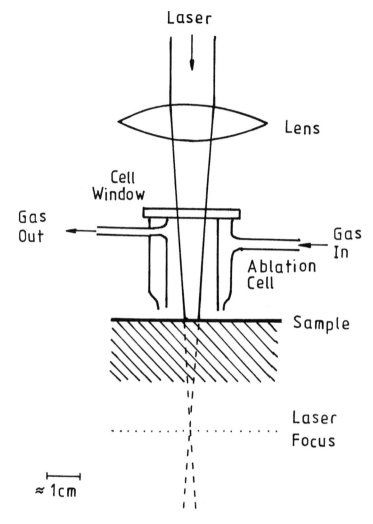

**Figure 4.32** Schematic of an open-ended ablation cell, positioned with a gap of 0.5–1 mm above the sample. The sample is fixed on an x-y-z translation stage, and the whole assembly is located in a gas-tight housing (neither shown).[91] Reproduced with the permission of the American Chemical Society.

was rotated with a stepper motor. The film of solution that adhered to the wheel then passed through the laser beam as the wheel rotated. The design of the rotating graphite disc for analysis of small amounts of solutions is reminiscent of the old technique called "rotrode" used in arc/spark spectrography.

The efficiency of the cell design, the transfer line, and the entrainment into the IC plasma was studied by Arrowsmith and coworkers.[225,226] Figure 4.34 shows the results. The efficiency of the system is determined by convolution of

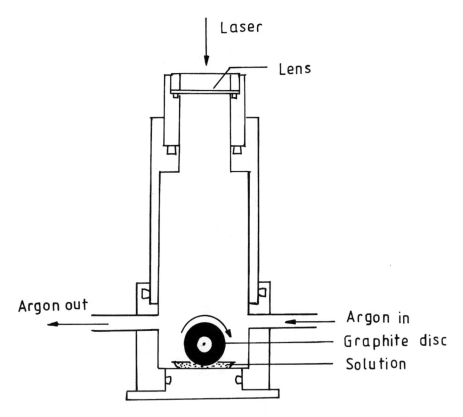

**Figure 4.33** Schematic diagram of a laser vaporization cell for analysis of small volumes of solutions.[86] Reproduced with permission of Elsevier Science.

three particle size functions: the size distribution of particles in the primary source, the transfer function of the cell and tube, and the response function of the secondary source. The cell and tube need transport particles only up to some size limit, corresponding to the fall-off in the response function of the secondary source. An ideal situation is shown in Figure 4.34b. A complete overlap of the primary source distribution, the transfer efficiency of the cell and tubing, and the response of the secondary source is demonstrated. The authors tried to explain how to reach this goal. Figure 4.35 shows an example for the calculated transport efficiency as a function of particle diameter for various flow rates, tube diameters, and lengths.

### 4.5.3.3 Spectrometers

Figure 4.30 is a schematic diagram of the main components of an ICP mass spectrometer with a quadrupole RF head. Additional main components of the instrument include the ICP torch, an interface system (consisting of a sampling

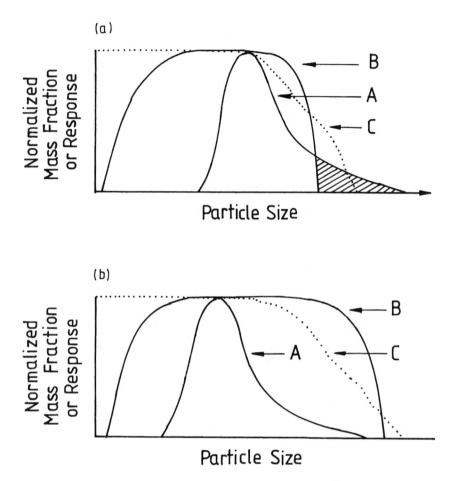

**Figure 4.34**  The overall efficiency of the system represented by convolution of particle size functions for A, the primary source, B, the transfer function of the cell and tube, and C, the response of the secondary source. Cases of partial and complete overlap of the primary source distribution are shown in (**a**) and (**b**), respectively.[225] Reproduced with permission of the author.

cone, a differentially pumped zone, and a skimming cone), ion lenses, and a detector coupled with the spectrometer head. In principle, the ICP torch is identical to that used for ICP-AES. However, grounding of the load coil and the interface system require unique considerations and designs. Critical parameters of effects of operating parameters on analyte signals in ICP-MS are the injector gas flow rate and the forward power. The injector flow along the axis of the torch is extracted by the aperture of the sampling cone, made from a metal of high conductivity such as nickel or platinum. Behind the extraction (or sampling) aperture, the gas pressure is maintained at about 5 mbar. The skimmer cone is usually more sharply angled than the extraction cone and machined with

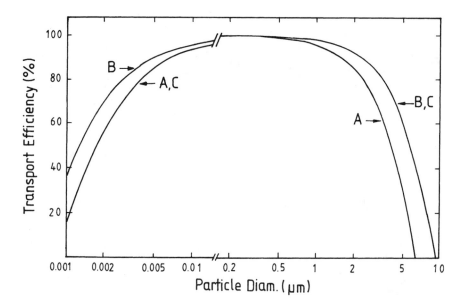

**Figure 4.35** Calculated transport efficiency for Mo particles (density = 10.2 g cm⁻³) for various flow rates, tube inner diameter, and length (A: 1.5 L/min, 4.5 mm, 80 cm; B: 1.5 L/min, 4.5 mm, 40 cm and 3.0 L/min, 4.5 mm, 80 cm; C: 1.5 L/min, 2.25 mm, 80 cm).[226] Reproduced with permission of the American Chemical Society.

a sharp lip about 5 μm wide. Beyond the skimmer, the extracted gas enters a region where the pressure is low enough to ensure that the mean free path becomes longer than the system dimensions. The form of the ion lens system is to some extent dictated by the nature of the vacuum system used. The function of the ion lens is to form as many ions as possible from the cloud at the rear of the skimmer into an axial beam of circular cross section at the entrance to the quadrupole mass analyzer. A quadrupole mass analyzer operates as a filter, along the axis of which a stable ion path exists for ions of only one mass. Quadrupoles provide only limited mass resolution. Some adjustment permits the use of higher resolution, if needed, to overcome a particular interference problem, and complete resolution of peaks at half–mass unit intervals from doubly charged ions is possible. The full advantage of the sensitivity of the technique can be exploited only by counting individual ions. Fast response can be obtained only by using electron multiplier detectors, and for most purposes continuous dynode channel electron multipliers are used. Despite the fact that almost all work in ICP-MS has been conducted using quadrupole mass spectrometers, certain important spectral interference problems can be eliminated only by the use of high-resolution sector-type mass spectrometers (Figure 4.36[227]). In sector fields mass spectrometry, ions have to be accelerated to a kinetic energy of a few kiloelectron volts prior to entering the analyzer. In Figure 4.37, a plasma interface concept is shown, developed by Gießmann und Greb.[227]

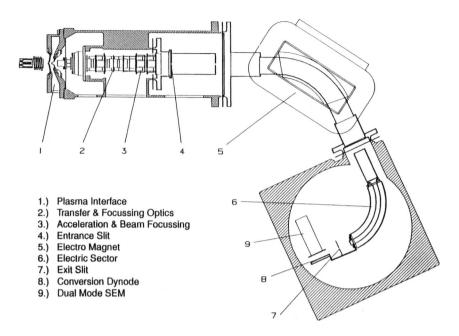

1.) Plasma Interface
2.) Transfer & Focussing Optics
3.) Acceleration & Beam Focussing
4.) Entrance Slit
5.) Electro Magnet
6.) Electric Sector
7.) Exit Slit
8.) Conversion Dynode
9.) Dual Mode SEM

**Figure 4.36** Schematic diagram of an ICP-MS, high-resolution type.

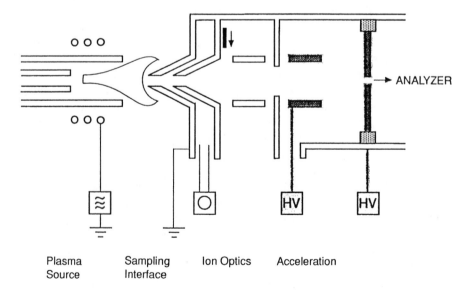

Plasma     Sampling     Ion Optics     Acceleration
Source     Interface

**Figure 4.37** Schematic diagram of the interface system between plasma source and analyzer of an HR-ICP-MS.

The plasma torch, sampler, and skimmer remain at ground potential. The acceleration of the ions to 8 kV takes place in the third vacuum chamber at a pressure $< 10^{-5}$ mbar. The double focusing sector field analyzer (Figure 4.36) is based on a reverse Nier-Johnson geometry. The mass dispersive magnetic field is located before the toroidal electric sector. Behind the exit slit the ions hit a conversion dynode; the electrons produced are amplified in a special two-stage electron multiplier (SEM).

### 4.5.4 Methods of Qualitative, Semiquantitative, and Quantitative Analysis

Some of the main points to be considered when carrying out analytical measurements with an ICP-MS are

1. Instrumental settings
   - RF power and gas flows of the ICP-MS
   - Laser parameters and optical arrangements
   - Cell design and transfer line
2. Spectral overlaps
   - Isobaric (interelement) spectral overlaps
   - Background spectral features
   - Oxide, hydroxide, and doubly charged species
   - Matrix-induced spectral overlaps
3. Calibration curve quality
   - Linearity
   - Dynamic range
   - Accuracy and precision
   - Limits of detection

Whereas instrument settings in defined ranges are given using a commercial LA-ICP-MS instrument, spectral overlaps have to be taken into account. In a sense, the periodic table is the table for ICP-MS, and therefore, spectral simplicity is expected, but spectral overlaps do exist in ICP-MS.

The distribution of ionization energies among the elements for singly and double charged ions are presented by Date and Gray.[228,p.18] Isobaric spectral overlaps occur, for example between $^{48}$Ti and $^{48}$Ca. In this case, the overlap can be corrected by measuring the signal for $^{44}$Ca and subtracting 0.09 of its measured count off the $^{48}$Ti. In a similar manner, the $^{58}$Fe overlap on $^{58}$Ni can be corrected by first measuring $^{56}$Fe, and the $^{64}$Ni overlap on $^{64}$Zn can be corrected by measuring $^{60}$Ni and subtracting the proportionate amount of the signal off the desired mass.[229]

The basic background species of ICP-MS consist of the species that are observed when water is nebulized into an argon ICP. The importance of the dry sample aerosol produced in LA-ICP-MS is that essentially all solvent-based molecular features are removed, considerably simplifying the background spectrum. This is not the case when using the liquid-solid calibration.

A number of spectral overlaps can be viewed as matrix-induced features. The oxide of a matrix element may cause a problem, for example, trace Cu determinations are difficult if Ti is present because $^{47}Ti^{16}O$ interferes with $^{63}Cu$, and $^{49}Ti^{16}O$ interferes with $^{65}Cu$. Ions such as $Cl^-$ and $SO_4^{2-}$ can generate numerous background molecular species. Argon itself combines with many species to form $ArCl^+$, $ArS^+$, and $ArP^+$. It should be noted that Ar seems to combine with just about any matrix element.

One of the capabilities of the multi-element method LA-ICP-MS is the so-called semiquantitative analysis for rapid characterization of the composition of samples. Although two ways of operation can be used, rapid spectral screening and peak hopping, the first way is the most common. The spectral scanning covers the entire mass range from mass-to-charge ratio ($m/z$) 1 to about 240 in approximately 30 s. Peak hopping to preprogrammed positions serves for identification of the presence of an element.

The mass/response relationship has been found to be applicable to a wide range of matrices, and can therefore be used in conjunction with an internal standard isotope to generate standardless semiquantitative analyses. The values obtained are usually within a factor of two to three of those obtained quantitatively with a set of standards, which may need to contain as many as 60 elements in total. Better accuracies with relative elemental responses of about 20, 30, and 50% have been published by Hager[230] and Broadhead et al.[231]

LA-ICP-MS can also be used for quantitative analysis. The accuracy that can be achieved in quantitation depends, to a large extent, on the availability of external standards and on the ability to perform internal standardization. Calibration curves can be established for either continuous or discrete pulse sampling. When good calibration standards are available, accuracy is often limited only by the analytical procedure. It is now typically within 1–10% (see application, Table 4.11). If certified reference materials are not available and matrix-matched standards cannot be produced, liquid-solid calibration is recommended as proposed and tested for LA-ICP-AES in 1989 by Thompson et al.[190] and Moenke-Blankenburg et al.[39a] Chenery and Cook[92] proved in 1993 the dual gas flow system shown in Figure 4.38 for simultaneous introduction of laser-ablated solids and nebulized aqueous solutions for calibration. Cromwell and Arrowsmith[91] obtained improved analytical results also, by calibration with aqueous solution standards using two-channel sample introduction to give equivalent ICP conditions for ablated material and a solution standard. They were using an adapter for mixed-sample introduction as shown in Figure 4.39. The adapter was inserted into the end of the torch.

### 4.5.5 Application

The fields of application have been reviewed by Moenke-Blankenburg[28] through 1992/93 and by Darke and Tyson[118,119] through 1993/94. Examples of up-to-date applications are summarized in Tables 4.11 and 4.12.

**Table 4.11** Application of LA-ICP-MS in Geological Sciences and Related Fields

| Elements | Matrices | Object of analysis | Ranges of concentrations | LOD (ppb) | RSD (%) | References |
|---|---|---|---|---|---|---|
| Li, Rb, Cs, Ti, Sc, V, Cr, Ni, Nb, Ta, Y, Th, U, REE et al. | Biotite, Zircon, Cordierite, Garnet, Apatite, Feldspar, Monazite | Study of abundance and distribution of trace elements in minerals | Trace | | Confidence intervals given | Bea et al.[49] |
| REE: La, Ce, Nd, Sm, Eu, Dy, Er, Yb | Monazite nodules | Single mineral grains; comparison with EPMA | 100 ppm level at 4 μm dia. crater | | 10 | Chenery, Cook[92] |
| 15 REE | IAEA soil-7 | Precision, accuracy; calibration | 0.4–20 ppm | 50–180 | 2.6–7.8 | Cousin, Magyar[50] |
| 23 elements | $U_3O_8$ CRM | Analysis of raw materials | Trace | 10–1000 | 8 | Crain, Gallimore[51] |
| Rb, Sr, Y, Zr, Nb | Amphibole, basanitic glass | Comparison SXRFM/ LA-ICP-MS | Major, minor | ppm level | 10–15 | Dalpé, Baker[232] |
| 30 elements | Chinese reference soils, GSS1-8 | Rapid multielement analysis | Major, minor, trace | ppm level 0.1–50 | Moderate | Durrant, Ward[233] |
| 30 elements | Chinese reference soils, GSS1-8 | Rapid multielement analysis | Major, minor, trace | ppm level: 0.1–50 | Moderate | Durrant, Ward[233] |
| Sc, V, Ni, Ga, Y, Zr, Hf, REE | Garnet | Comparison LA-ICP-MS to PIXE | Trace | | 8 | Fedorowich et al.[234] |
| $^{207}Pb/^{206}Pb$, | Zircon | Geochronology | Ratios | ppm level | 1–2 | Feng et al.[54] |

*(continued)*

**Table 4.11** Application of LA-ICP-MS in Geological Sciences and Related Fields (*Continued*)

| Elements | Matrices | Object of analysis | Ranges of concentrations | LOD (ppb) | RSD (%) | References |
|---|---|---|---|---|---|---|
| 26 el.-oxides | Plumboan, chrichtonite | Comparison EPMA/LA-ICP-MS; Senaite identification | Major, minor | | | Foord et al.[235] |
| U/Pb | Pitchblende, Zircon | Geochronology | Ratios | | 0.1–2 | Fryer et al.[55] |
| Pt, REE | Ferromanganese crusts | In situ analysis | Trace | | 4–5 | Garbe-Schönberg, McMurtry[56] |
| 13 elements | Dolomite, Zircon; GSJ | Distribution analysis | Major, minor, trace | | | Imai, Yamamoto[236] |
| 26 elements | Silicate rocks; ref. stand. | Rapid analysis | Major, minor, trace | ppb to ppm | ~10 | Jarvis, Williams[60] |
| Zr, Nb, Ta | Clinopyroxene, garnet, rutile, glass | Determination of partition coefficients; comparison PIXE | Trace | | | Jenner et al.[94] |
| REE | Dolomitized carbonates | Variability in REE contents | Trace | | | Kontak, Jackson[237] |
| 49 elements | Rocks, soils | Lithium borate fusions | Major, minor, trace | 10 | 5 | Lichte[238] |
| Li, Be, Sc, V, Cr, Co, Ni, Rb, Sr, Y, Zr, Ba, REE, Hg, Th, U/Pb, | Zircon, calcite, dolomite, rocks, standards | Isotopic and trace element analysis; geochronology | Trace | 10 (Ba, REE) to 2000 (Hg) | 2–5 | Ludden et al.[239] |
| Cu, Zn, Sn, Pb, Mg, Mn, Fe, Sr, Ba, U | CaCO3, limestone, GSJ JLs-1, BCS CRM 393 | Semiquantitative, quantitative analysis | Trace, ppm range | | ~10 | Pearce et al.[63] |

| Elements | Material | Purpose | Type | | | Reference |
|---|---|---|---|---|---|---|
| Hf, Y, REE, Th, U | Zircons, zircon sands, Zr concentrates; CRMs from UK, S. Africa, Australia | Test of the validity of LA-ICP-MS to analyze CRMs | Minor, trace | | 1.7 to ~50 | Perkins et al.[65] |
| Li, B, Mg, K, Sc, V, Cr, Mn, Co, Ni, Cu, Zn, Ga, Rb, Sr, Y, Zr, Nb, Mo, Ag, Cd | Glass CRMs; NIST SRM612, NIST SRM610, P&H DLH6, P&H DLH7, P&H DLH8 | Test of a new series of glass standards | | | 2–3 | Raith et al.[67] |
| Mg, Al, Mn, Fe, Zn, Sr, Ba, La, Pb | Marble crusts | Study of weathering layers | Minor, trace | | 4.6–9.4 | Ulens et al.[69] |
| Cd, Zn, Te, Bi, Cr, Pb | Natural gold | Fingerprints; different origin | Trace | | | Watling et al.[72] |
| Ce, Fe | Volcanic glass | Tephrochronological study | Trace | | <20 | Westgate et al.[74] |
| REE, Hf, Ta, W | Silicate rocks | Multi-element determination | Minor, trace | <100 | <10 | Williams, Jarvis[75] |

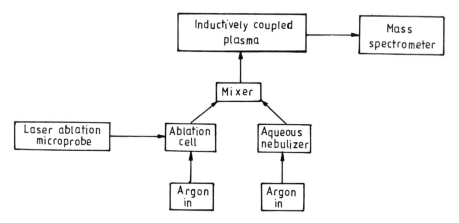

**Figure 4.38** Block diagram of a dual gas-flow sample introduction system.[92] Reproduced with permission of Elsevier Science.

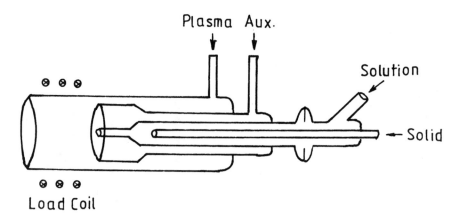

**Figure 4.39** Schematic of an adapter for mixed-sample introduction into an ICP. The adapter is inserted into the end of the torch and attached by a ground-glass seal with a spring clamp (clamp not shown).[91] Reproduced with permission of the American Chemical Society.

**Table 4.12**  Application of LA-ICP-MS in Biology

| Elements | Matrices | Object of analysis | Ranges of concentration | LOD (ppb) | RSD (%) | References |
|---|---|---|---|---|---|---|
| Cd,Pb, Cu | Calcified biological structures | Trace metal contaminants | Trace | | | Evans et al.[52] |
| Pb, Cu, Zn, Sr | Teeth of walrus | History of metal exposure | Trace | | | Evans et al.[53] |
| Mg, Al, Ca, Cr, Mn, Fe, Co, Ni, Cu, Zn, Sr, Cd, Ba, TL, PB, Bi, U | Trees | Annual growth rings of trees | Trace | 10 (Mn, Cd) to 4000 (Ca) | | Hoffmann et al.[59] |
| Ca, Sr, Zn, Pb, Co, Cu | Mammalian teeth | Tooth type and trace metals | Trace | | | Outridge, Evans[85] |
| Sr, Bi, Rb | Fish scale, rat kidney, pig femur | Distribution analysis | Trace | | | Wang et al.[71] |

# References

1. Maiman, T.H. *Phys. Rev. Lett.* **1960,** *4,* 564; *Nature (London),* **1960,** *187,* 493.

2. Anonymous, *Chem. Eng. News* **1962,** *40,* 52.

3. Brech, F.; Cross, L. *Appl. Spectrosc.* **1962,** *16,* 59.

4. Debras-Guedon, J.; Liodec, N. *C. R. Acad. Sci.* **1963,** *257,* 336.

5. Runge, E.F.; Bryan, F.R.; Minck, R.W. *Can. J. Spectrosc.* **1964,** *9,* 5.

6. Ferguson, H.I.S.; Mentall, J.E.; Nicholis, R.W. *Nature (London)* **1964,** *204,* 1295.

7. Berndt, M.; Krause, H.; Moenke-Blankenburg, L.; Moenke, H. *Jenaer Jahrb.* **1965,** 45.

8. Debras-Guedon, J.; Liodec, N.; Vilnat, J. *J. Math. Phys. Appl.* **1965,** *16,* 155.

9. Hagenah, W.D. *Z. Angew. Math. Phys.* **1965,** *16,* 130.

10. Jankewitsch, W.F.; Besrutschko, J.W. *Zavod. Lab.* **1965,** *30,* 628.

11. Karyakin, A.V.; Achmanova, M.V.; Kaigorodov, V.A. *Zh. Anal. Khim.* **1965,** *20,* 145.

12. Moenke, H.; Moenke-Blankenburg, L. *Einführung in die Laser-Mikro-Emissionsspektralanalyse;* Akademische Verlagsgesellschaft Geest & Portig KG: Leipzig, Germany, 1966.

13. Honig, R.E.; Woolston, J.R. *Appl. Phys. Lett.* **1963,** *2,* 138.

14. Honig, R.E. *Appl. Phys. Lett.* **1963,** *3,* 8.

15. Giory, F.A.; McKenzie, L.A.; McKinney, E.J. *Appl. Phys. Lett.* **1963,** *3,* 25.

16. Isenor, N.R. *Appl. Phys. Lett.* **1963,** *4,* 152.

17. Isenor, N.R. *Can. J. Chem.* **1964,** *42,* 1413.

18. Berkowitz, J.; Chupka, W.A. *J. Chem. Phys.* **1964,** *40,* 2735.

19. Lincoln, K.A. *Anal. Chem.* **1965,** *37,* 541.

20. Isenor, N.R. *J. Appl. Phys.* **1965,** *36,* 316.

21. Knox, B.E.; Vastola, F.J. *Chem. Eng. News* **1966,** *44,* 48.

22. Anisimov, S.I.; Bronch-Brujevich, A.M.; Elyashevich, M.A.; Imas, Y.A.; Pavlenko, N.A.; Romanov, G.S. *Zh. Tekh. Fiz.* **1966,** *36,* 1273.

23. Bernal, E.; Levine, L.P.; Ready, J.F. *Rev. Sci. Instrum.* **1966,** *37,* 938.

24. Eloy, J.-F.; Dumas, J.L. *Rev. Methods Phys. Anal.* **1966,** *2,* 251.

25. Fenner, N.C.; Daly, N.R. *Rev. Sci. Instrum.* **1966,** *37,* 1068.

26. Fenner, N.C. *Phys. Lett.* **1966,** *22,* 421.

27. Langer, P.; Tonon, G.; Floux, F.; Ducauze, A. *IEEE J. Quantum Electron.* **1966,** *2,* 499.

28. Moenke-Blankenburg, L. *Spectrochim. Acta Rev.* **1993,** *15,* 1.

29. Thompson, M.; Goulter, J.E.; Sieper, F. *Jena Rev.* **1981,** *26,* 202; *Analyst* **1981,** *106,* 32.

30. Carr, J.W.; Horlick, G. *Spectrochim. Acta* **1982,** *37B,* 1.

31. Ishizuka, T.; Uwamino, Y. *Spectrochim. Acta* **1983,** *38B,* 519.

32. Dittrich, K.; Niebergall, K.; Wennrich, R. *Fresenius Z. Anal. Chem.* **1987,** *328,* 330.

33. Günther, D.; Gäckle, M. *Z. Chem.* **1988,** *28,* 227 and 258.

34. Thompson, M.; Flint, C.D.; Chenery, S. *J. Anal. Atom. Spectrom.* **1988**, *3*, 1133.

35. Chenery, S.; Thompson, M.; Timmins, K. *Anal. Proc.* **1988**, *25*, 68.

36. Mochizuki, T.; Sakashita, A.; Seno, H.; Iwata, H. *Bunseki Kagaku* **1988**, *37*, 12 and T 109.

37. Wennrich, R.; Niebergall, K.; Zwanziger, H.; Just, G. *Isotopenpraxis* **1988**, *24*, 97.

38. Günther, D.; Gäckle, M.; Moenke-Blankenburg, L.; Abicht, H.-P. *Silikattechnik* **1990**, *41*, 10.

39a. Moenke-Blankenburg, L. Processes of Laser Ablation and Vapour Transport to the ICP; Plenary Lecture; *Third Surrey Conference on Plasma Source Mass Spectrometry*, University of Surrey, Guildford, July 16–19, 1989 (see reference 39b).

39b. Moenke-Blankenburg, L.; Gäckle, M.; Günther, D.; Kammel, J. Plasma Source Mass Spectrometry; Special Publication No. 85; Royal Society of Chemistry; Cambridge, UK, 1990, pp. 1–17.

40. Thompson, M.; Chenery, S.; Brett, L. *J. Anal. Atom. Spectrom.* **1990**, *5*, 49.

41. Dittrich, K.; Mohamed, I.; Nyuyen, H.T.; Niebergall, K.; Pfeifer, M.; Wennrich, R. *Fresenius Z. Anal. Chem.* **1990**, *337*, 546.

42. Günther, D.; Gäckle, M. Ph.D. Thesis, Martin-Luther-Universität Halle-Wittenberg, Halle, Germany, 1990.

43. Mochizuki, T.; Sakashita, A.; Tsuji, T.; Iwata, H.; Ishibashi, Y.; Gunju, N. *Anal. Sci.* **1991**, *7*, 479.

44. Moenke-Blankenburg, L.; Schumann, T.; Günther, D; Kuß, H.-M.; Paul, M. *J. Anal. Atom. Spectrom.* **1992**, *7*, 251.

45. Moenke-Blankenburg, L.; Günther, D. *Chem. Geol.* **1992**, *95*, 85.

46. Moenke-Blankenburg, L.; Schumann, T.; Nölte, J. *J. Anal. Atom. Spectrom.* **1994**, *9*, 1059.

47. Moenke-Blankenburg, L.; Kammel, J.; Schumann, T. *Microchem. J.* **1994**, *50*, 374.

48. Alimpeev, S.S.; Belov, M.E.; Nikiforov, S.M. *Anal. Chem.* **1993**, *65*, 3194.

49. Bea, F.; Pereira, M.D.; Stroh, A. *Chem. Geol.* **1994**, *117*, 291.

50. Cousin, H.; Magyar, B. *Mikrochim. Acta.* **1994**, *113*, 313.

51. Crain, J.S.; Gallimore, D.L. *J. Anal. Atom. Spectrom.* **1992**, *7*, 605.

52. Evans, R.D.; Outridge, P.M. *J. Anal. Atom. Spectrom.* **1994**, *9*, 985.

53. Evans, R.D.; Richner, P.; Outridge, P.M. *Arch. Environ. Contam. Toxicol.* **1995**, *28*, 55.

54. Feng, R.; Machado, N.; Ludden, J. *Geochim. Cosmochim. Acta* **1993**, *57*, 3479.

55. Fryer, B.J.; Jackson, S.E.; Longerich, H.P. *Chem. Geol.* **1993**, *109*, 1.

56. Garbe-Schönberg, C.-D.; McMurtry, G.M. *Fresenius J. Anal. Chem.* **1994**, *350*, 264.

57. Günther, D.; Longerich, H.P.; Forsythe, L.; Jackson, S.E. *Amerik. Lab.* **1995**, *June*, p. 24.

58. Günther, D.; Longerich, H.P.; Jackson, S.E. *J. Can. Appl. Spectrosc.* **1995**, *40*, 1.

59. Hoffmann, E.; Lüdke, C.; Scholze, H.; Stephanowitz, H. *Fresenius J. Anal. Chem.* **1994**, *350*, 253.

60. Jarvis, K.E.; Williams, J.G. *Chem. Geol.* **1993**, *106*, 251.

61. Kogan, V.V.; Hinds, M.W.; Ramendik, G.I. *Spectrochim. Acta* **1994**, *49B*, 333.

62. Lüdke, C.; Hoffmann, E.; Skole, J. *Fresenius J. Anal. Chem.* **1994**, *350*, 272.

63. Pearce, N.J.G.; Perkins, W.T.; Fuge, R. *J. Anal. Atom. Spectrom.* **1992,** *7,* 595.

64. Perkins, W.T.; Pearce, N.J.G.; Fuge, R. *J. Anal. Atom. Spectrom.* **1992,** *7,* 611.

65. Raith, A.; Hutton, R.C. *Fresenius J. Anal. Chem.* **1994,** *350,* 242.

66. Richner, R.; Evans, D. *Atom. Spectrosc.* **1993,** *14,* 157.

67. Raith, A.; Godfrey, J.; Hutton, R.C. *Fresenius J. Anal. Chem.* **1996,** *354,* 163.

68. Scholze, H.; Stephanowitz, H.; Hoffmann, E.; Skole, J. *Fresenius J. Anal. Chem.* **1994,** *350,* 247.

69. Ulens, K.; Moens, L.; Dams, R.; Van Wickel, S.; Vendevelde, L. *J. Anal. Atom. Spectrom.* **1994,** *9,* 1243.

70. Walder, A.J.; Abell, I.D.; Platzner, I. *Spectrochim. Acta.* **1993,** *48B,* 397.

71. Wang, S.; Brown, R.; Gray, D.J. *Appl. Spectrosc.* **1994,** *48,* 1321.

72. Watling, R.J.; Herbert, H.K.; Delev, D.; Abell, I.D. *Spectrochim. Acta* **1994,** *49B,* 205.

73. van de Weijer, P.; Baeten, W.L.M.; Bekkers, M.H.J.; Vullings, P.J.M.G. *J. Anal. Atom. Spectrom.* **1992,** *7,* 599.

74. Westgate, J.A.; Perkins, W.T.; Fuge, R.; Pearce, N.J.G.; Wintle, A.G. *Appl. Geochem.* **1994,** *9,* 323.

75. Williams, J.G.; Jarvis, K.E. *J. Anal. Atom. Spectrom.* **1993,** *8,* 25.

76. Yasuhara, H.; Okano, T.; Matsumura, Y. *Analyst* **1992,** *117,* 395.

77. Anderson, D.R.; McLeod, C.W.; Smith, T.A. *J. Anal. Atom. Spectrom.* **1994,** *9,* 67.

78. Baldwin, D.P.; Zamzow, D.S.; D'Silva, A.P. *Anal. Chem.* **1994,** *66,* 1911.

79. Nölte, J.; Moenke-Blankenburg, L.; Schumann, T. *Fresenius J. Anal. Chem.* **1994,** *349,* 131.

80. Nölte, J.; Schöppenthau, J.; Dunemann, L.; Schumann, T.; Moenke-Blankenburg, L. *J. Anal. Atom. Spectrom.* **1995,** *10,* 655.

81. Moenke-Blankenburg, L. In *Laser Ionization Mass Analysis;* Vertes, A.; Gijbels, R.; Adams, F., Eds.; Wiley & Sons: New York, 1993; pp. 433–452.

82. Wisbrun, R.; Schechter, I.; Niessner, R.; Schröder, H.; Kompa, K.L. *Anal. Chem.* **1994,** *66,* 2964.

83. Borthwick, S.; Clark, A.; Ledingham, K.W.D.; Singhal, R.P. In *AIP Conf. Proc. 288, Laser Ablation: Mechanisms and Applications II, 1993;* Miller, J.C.; Geohegan, D.B., Eds.; (American Institute of Physics Press, New York): 1994; pp. 165–171.

84. Cleatham, M.M.; White, W.M. In Proc. 1994 Winter Conf. Plasma Spectrochem., San Diego, CA, U.S.A., WP41, p. 177, Barnes, R.M., Ed., Amherst, MA, U.S.A., January 1994.

85. Outridge, P.M.; Evans, R.D. *J. Anal. Atom. Spectrom.,* submitted (1996).

86. Prabhu, R.K.; Vijayalakshmi, S.; Mahalingam, T.R.; Viswanathan, K.S.; Mathews, C.K. *J. Anal. Atom. Spectrom.* **1993,** *8,* 565.

87. Tang, K.; Allman, S.L.; Chen, C.H. In *AIP Conf. Proc. 288, Laser Ablation: Mechanisms and Applications II, 1993;* Miller, J.C.; Geohegan, D.B., Eds.; AIP Press: 1994; pp. 475–482.

88. Hastie, J.W.; Bonnell, D.W.; Paul, A.J.; Schenck, P.K. *Mat. Res. Soc. Symp. Proc.* **1993,** *285,* 39.

89. Zamzow, D.S.; Baldwin, D.P., Weeks, S.J.; Bajic, S.J.; D'Silva, A.P. *Environ. Sci. Technol.* **1994,** *28,* 352.

90. Bickel, G.A.; McRae, G.A.; Green, L.W. In *AIP Conf. Proc. 288, Laser Ablation: Mechanisms*

*and Applications II, 1993;* Miller, J.C.; Geohegan, D.B., Eds.; American Institute of Physics Press (New York): 1994; pp. 178–182.

91. Cromwell, E.F.; Arrowsmith, P. *Anal. Chem.* **1995**, *67,* 131.

92. Chenery, S.; Cook, J.M. *J. Anal. Atom. Spectrochem.* **1993**, *8,* 299.

93. Longerich, H.P.; Günther, D.; Jackson, S.E. *Fresenius J. Anal. Chem.* **1996**, *355,* 538.

94. Jenner, G.A.; Foley, S.F.; Jackson, S.E.; Green, T.H.; Fryer, B.J.; Longerich, H.P. *Geochim. Cosmochim. Acta* **1994**, *58,* 5099.

95. Nicholls, M.; Mills, D.J.; Newton, I.A.; Turner, P.J. *Proc. 1994 Winter Conf. Plasma Spectrochem.,* San Diego, CA, U.S.A., W51, p. 187, Barnes, R.M., Ed., Amherst, MA, U.S.A., January 1994.

96. Lin, S.; Peng, C. *J. Anal. Atom. Spectrom.* **1990**, *5,* 509.

97. Lin, S.; Zheng, C.; Chen, Y.; Wu, C. *Proc. 1994 Winter Conf. Plasma Spectrochem.,* San Diego, CA, U.S.A., WP48, p. 184.

98. Kagawa, K.; Matsuda, Y.; Zokoi, S.; Nakajima, S. *J. Anal. Atom. Spectrom.* **1988**, *3,* 415.

99. Champeaux, C.; Marchet, P.; Catherinot, A. In *AIP Conf. Proc. 288, Laser Ablation: Mechanisms and Applications II, 1993;* pp. 433–438, Miller, J.C., Geohegan, D.B., Eds; AIP Press, New York.

100. Fukumura, H.; Mibuka, N.; Eura, S.; Masuhara, H.; Nishi, N. *J. Phys. Chem.* **1993,** *97,* 13761.

101. Solis, J.; Vega, F.; Afonso, C.N.; Georgiou, E.; Charalambidis, D.; Fotakis, C. *J. Appl. Phys.* **1993**, *74,* 4271.

102. André, N.; Geertsen, C.; Lacour, J.-L.; Mauchien, P.; Sjöström, S. *Spectrochim. Acta* **1994,** *49B,* 1363.

103. Sabsabi, M.; Cielo, P.; Boily, S.; Chaker, M. *Proc. SPIE* **1993,** *2069,* 190.

104. Sabsabi, M.; Cielo, P. *Proc. 1994 Winter Conf. Plasma Spectrochem.,* San Diego, CA, U.S.A., WP61, p. 199.

105. Russo, R.E.; Chan, W.T.; Bryant, M.F.; Kinard, W.F. *J. Anal. Atom. Spectrom.* **1995**, *10,* 295.

106. Fernandes, A.; Mao, X.L.; Chan, W.T.; Channon, M.A.; Russo, R.E. *Anal. Chem.* **1995,** *67,* 2444.

107. Maitland, A.; Dunn, M.H. *Laser Physics;* North Holland Publishing: Amsterdam, 1969.

108. Bögershausen, W.; Vesper, R. *Spectrochim. Acta* **1969,** *24B,* 103.

109. Ready, J.F. *Effects of High-Power Laser Radiation;* Academic: New York, 1971.

110. Krokhin, O.N. In *Laser Handbook;* Arech, F.T.; Schulz-du Bois, E.O., Eds.; North-Holland Publishing: Amsterdam, 1972; Vol. 2, p. 1371.

111. Hughes, T.P. *Plasma and Laser Light;* Adam Hilger: Bristol, UK; John Wiley & Sons: New York, 1975.

112. Sacchi, C.A.; Svelto, O. In *Analytical Laser Spectroscopy;* Omenetto, N., Ed.; John Wiley & Sons, New York, 1979; pp. 1–46.

113. Eloy, J.F. *Les Lasers de Puissance;* Masson Editeur: Paris, 1985.

114. Piepmeier, E.H. *Analytical Applications of Lasers;* John Wiley & Sons: New York, 1986.

115. Cremers, D.A.; Radziemski, L.J. In *Laser Spectroscopy and its Applications;* Radziemski, L.J.; Solarz, R.W.; Paisner, J.A., Eds.; Marcel Dekker: New York, 1987.

116. Lubman, D.M. *Lasers and Mass Spectroscopy;* Oxford University Press: New York, 1989.

117. Moenke-Blankenburg, L. *Laser Microanalysis;* Winefordner, J.D.; Kolthoff, I.M., Eds.; John Wiley & Sons: New York, 1989.

118. Darke, S.A.; Tyson, J.F. *J. Anal. Atom. Spectrom.* **1993,** *8,* 145.

119. Darke, S.A.; Tyson, J.F. *Microchem. J.* **1994,** *50,* 310.

120. Mazhukin, V.I.; Gusev, I.V.; Smurov, L.; Flamant, G. *Microchem. J.* **1994,** *50,* 413.

121. Radziemski, L.J. *Microchem. J.* **1994,** *50,* 218.

122. Lee, Y.I.; Sneddon, J. *Microchem. J.* **1994,** *50,* 235.

123. Vertes, A.; Juhasz, P.; DeWolf, M.; Gijbels, R. *Scanning Microsc.* **1988,** *2,* 1853.

124. Vertes, A.; Juhasz, P.; DeWolf, M.; Gijbels, R. *Int. J. Mass Spectrom. Ion Proc.* **1989,** *94,* 63.

125. Vertes, A.; Gijbels, R.; Adams, F. *Mass Spectrom. Rev.* **1989,** *9,* 71.

126. Adams, F.; Vertes, A. *Fresenius J. Anal. Chem.* **1990,** *337,* 638.

127. Mermet, J.M. *Quim. Anal. (Barcelona)* **1989,** *8,* 385.

128. Balasz, L.; Gijbels, R.; Vertes, A. *Anal. Chem.* **1991,** *63,* 311.

129. Hess, P., Ed.; *Photoacoustic, Photothermal, and Photochemical Processes at Surfaces and in Thin Films;* Springer: Berlin, 1989; Chapters 3 and 4.

130. Moenke, H.; Moenke-Blankenburg, L. *Einführung in die Laser-Mikro-Emissionsspektralanalyse,* 2nd ed. in German; Akademische Verlagsgesellschaft Geest & Portig: Leipzig, Germany, 1968; 3rd ed. in russian; MIR: Moscow, Russia, 1968.

131. Moenke, H.; Moenke-Blankenburg, L. *Laser Micro-Spectrochemical Analysis;* Adam Hilger: London, 1973.

132. Scott, R.H.; Strasheim, A. In *Applied Atomic Spectroscopy;* Grove, E.L., Ed.; Plenum: New York, 1978; Vol. 1, Chapter 2, pp. 73–118.

133. Leis, F.; Sdorra, W.; Ko, J.-B.; Niemax, K. *Mikrochim. Acta* **1989,** *2,* 85.

134. Sdorra, W.; Quentmeier, A.; Niemax, K. *Mirochim. Acta* **1989,** *2,* 201.

135. Radziemski, L.J.; Cremers, D.A. In *Laser-Induced Plasmas and Applications;* Radziemski, L.J.; Cremers, D.A., Eds.; Marcel Dekker: New York, 1989; Chapter 7, pp. 295–325.

136. Kim, Y.W. In *Laser-Induced Plasmas and Applications;* Radziemski, L.J.; Cremers, D.A., Eds.; Marcel Dekker: New York, 1989; Chapter 8, pp. 327–346.

137. Beauchemin, D.; Le Blanc J.C.Y.; Peters, G.R.; Craig, J.M. *Anal. Chem.* **1992,** *64,* 442R.

138. Cates, M.C. *Proc. SPIE Int. Soc. Opt. Eng.* **1990,** *1279,* 102.

139. Niemax, K.; Sdorra, W. *Appl. Opt.* **1990,** *29,* 5000.

140. Quentmeier, A.; Sdorra, W.; Niemax, K. *Spectrochim. Acta* **1990,** *45B,* 537.

141. Sdorra, W.; Niemax, K. *Spectrochim. Acta* **1990,** *45B,* 1917.

142. Grant, K.J.; Paul, G.L.; O'Neill, J.A. *Anal. Sci.* **1990,** *44,* 1711.

143. Timmer, C.; Srivastava, S.K.; Hall, T.E.; Fucaloro, A.F. *J. Appl. Phys.* **1991,** *70,* 1888.

144. Cassini, M.; Harieth, M.A.; Palleschi, V.; Salvetti, A.; Singh, D.P.; Vaselli, M. *Laser Part. Beams* **1991,** *9,* 633.

145. Wisbrun, R.; Niessner, R.; Schröder, H. *Anal. Methods Instrum.* **1993,** *1,* 1.

146. Jowitt, R. *Proc. 38. Chemist's Conf.*, Scarborough, June 25–27, 1985, pp. 19–29.

147. Lorenzen, C.J.; Carlhoff, C.; Hahn, U.; Jogwich, M. *J. Anal. Atom. Spectrom.* **1992,** 7, 1029.

148. Piepmeier, E.H.; Osten, D.E. *Appl. Spectrosc.* **1971,** 25, 642.

149. Ownes, M.; Majidi, V. *Appl. Spectrosc.* **1991,** 45, 1463.

150. Talmi, Y.; Sieper, H.-P.; Moenke-Blankenburg, L.; Quillfeldt, W. *Abstracts of Papers*, Proc. 21st Colloquium Spectroscopicum Internationale, Cambridge; 1979, Abstract 40. The Burlington Press, Cambridge, UK, July 1979.

151. Talmi, Y.; Sieper, H.-P.; Moenke-Blankenburg, L. *Anal. Chim. Acta* **1981,** *127,* 71.

152. Han, M.Y.; Cremers, D.A. submitted to *Int. J. Environ. Anal. Chem.* (1992).

153. Lee, Y.-I.; Sneddon, J. *Analyst* **1994,** *119,* 1441.

154. Idzikowski, A.F.; Gadek, S. *Chem. Anal. (Warsaw)* **1994,** *39,* 67.

155. Moenke-Blankenburg, L. *Proceedings of CANAS '93 in Oberhof;* Dittrich, K., Ed.; University of Leipzig: Leipzig, Germany, 1993; pp. 165–180.

156. Abercrombie, F.N.; Silvester, M.D.; Stoute, G.S. Presented at the 28th Pittsburgh Conference, Cleveland, OH, 1977; paper 406; and *ICP Inf. Newsl.* **1977,** *2,* 309.

157. Salin, E.D.; Carr, J.W.; Horlick, G. Presented at the 30th Pittsburgh Conference, Cleveland, OH, 1979; paper 563.

158. Carr, J.W.; Horlick, G. Presented at the 31st Pittsburgh Conference, Atlantic City, NJ, 1980; paper 56.

159. Su, G.; Lin, S. *J. Anal. Atom. Spectrom.* **1988,** *3,* 841.

160. Liu, X.R.; Horlick, G. *Spectrochim. Acta* **1995,** *50B,* 537.

161. Liu, X.R.; Horlick, G. *J. Anal. Atom. Spectrom.* **1994,** *9,* 833.

162. Mitchell, P.G.; Sneddon, J.; Radziemski, L.J. *Appl. Spectrosc.* **1986.** *40,* 274.

163. Mitchell, P.G.; Sneddon, J.; Radziemski, L.J. *Appl. Spectrosc.* **1987,** *47,* 141.

164. Thompson, M.; Flint, C.D.; Chenery S.; Knight, K. *J. Anal. Atom. Spectrom.* **1992,** *7,* 1099.

165. Chenery, S.; Hunt, A.; Thompson, M. *J. Anal. Atom. Spectrom.* **1992,** *7,* 647.

166. Scott, R.H.; Fassel, V.A.; Kniseley, R.N.; Nixon, D.E. *Anal. Chem.* **1974,** *46,* 75.

167. Horlick, G. *Appl. Spectrosc.* **1976,** *30,* 113.

168. Dickinson, G.W.; Fassel, V.A. *Anal. Chem.* **1969,** *41,* 1021.

169. Runge, C.R.; Paschen, F. *Abh. K. Akad. Wiss. Berlin* **1902,** *1.*

170. Ebert, H. *Ann. Physik* **1889,** *38,* 489.

171. Fastie, W.G. *J.O.S.A.* **1952,** *42,* 641.

172. Seya, M. *Sci. Light* **1959,** *8,* 39.

173. Namioka, T. *J.O.S.A.* **1959,** *49,* 951.

174. Barnard, T.W.; Crockett, M.I.; Ivaldi, J.C.; Lundberg, P.L. *Anal. Chem.* **1993,** *65,* 1225.

175. Quillfeldt, W. *Fresenius J. Anal. Chem.* **1991,** *340,* 459.

176. Horlick, G. *Appl. Spectrosc.* **1976,** *30,* 113.

177. Lepla, K.C.; Horlick, G. *Appl. Spectrosc.* **1989,** *43,* 1187.

178. Lepla, K.C.; Horlick, G. *Appl. Spectrosc.* **1990**, *44*, 1259.

179. McGeorge, S.W.; Salin, E.P. *Prog. Anal. Spectrosc.* **1981**, *7*, 387.

180. Boumans, P.W.J.M. *J. Anal. Atom. Spectrom.* **1993**, *8*, 767.

181. Barnard, T.W.; Crockett, M.I.; Ivaldi, J.C.; Lundberg, P.L.; Yates, D.A.; Levine, P.A.; Sauer, D.J. *Anal. Chem.* **1993**, *65*, 1231.

182. Pilon, M.J.; Denton, M.B.; Schleicher, R.G.; Moran, P.M.; Smith, S.B., Jr. *Appl. Spectrosc.* **1990**, *44*, 1613.

183. Pang, H.; Wiederin, D.R.; Houk, R.S.; Yeung, E.S. *Chem.* **1991**, *63*, 390.

184. Darke, S.A.; Long, S.E.; Pickford, C.J.; Tyson, J.F. *J. Anal. Atom. Spectrom.* **1989**, *4*, 715.

185. Darke, S.A.; Long, S.E.; Pickford, C.J.; Tyson, J.F. *Anal. Proc.* **1989**, *26*, 159.

186. Darke, S.A.; Long, S.E.; Pickford, C.J.; Tyson, J.F. *Fresenius J. Anal. Chem.* **1990**, *337*, 284.

187. Mochizuki, T.; Sakashita, A.; Akiyoshi, T.; Iwata, H. *Anal. Sci.* **1989**, *5*, 535.

188. Richner, P.; Borer, M.W.; Brushwyler, K.R.; Hieftje, G.M. *Appl. Spectrosc.* **1990**, *44*, 1290.

189. Moenke-Blankenburg, L.; Pacher, M. *Acta Chim. Hung.* **1989**, *126*, 40.

190. Danzer, K.; Küchler, L. *Talanta* **1977**, *24*, 561.

191. Thompson, M.; Chenery, S.; Brett, L. *J. Anal. Atom. Spectrom.* **1989** *4*, 11.

192. Thiem, T.L.; Wolf, P.J. *Microchim. Acta* **1994**, *50*, 244.

193. Wilkinson, J.J.; Rankin, A.H.; Mulshaw, S.C.; Nolan, J.; Ramsay, M.H. *Geochim. Cosmochim. Acta* **1994**, *58*, 1133.

194. Rankin, A.H.; Herrington, R.J.; Ramsay, M.R; Coles, B.; Christoula, M.; Jones, E. *Proceedings of the 8th Quadrennial IAGOD Symposium 1994;* E. Schweizerbrthśche Verlagsbuchhandlung Nägele & Obermiller: Stuttgart, Germany; pp. 183–198.

195. Russo, R.E. *Appl.Spectrosc.* **1995**, *49*, 14A.

196. Shannon, M.A.; Mao, X.L.; Fernandez, A.; Chan, W.-T.; Russo, R.E. submitted to *Anal. Chem.* **1995;**

197. Mao, X.L.; Shannon, M.A.; Fernandez, A.J.; Russo, R.E. *Appl. Spectrosc.* **1995**, *49*, 1054.

198. Mao, X.L.; Chan, W.-T.; Shannon, M.A.; Russo, R.E. *J. Appl. Phys.* **1993**, *74*, 4915.

199. Geertsen, C.; Briand, A; Chartier, F.; Lacour, J.L.; Mauchien, P.; Sjöström, S.; Mermet, J.-M. *J. Anal. Atom. Spectrom.* **1994**, *9*, 17.

200. Hillenkamp, F.; Karras, M.; Beavis, R.C.; Chait, B.T. *Anal. Chem.* **1991**, *63*, 1193A.

201. *LIMS Reference and Citation Index;* University of Düsseldorf: Düsseldorf, Germany, 1991.

202. Lee, I.; Coutant, C.C.; Arakawa, E.T. Published by the American Institute of Physics, New York, **1994**, pp. 172–177.

203. Preuß, S.; Stuke, M. *Ber. Bunsenges. Phys. Chem.* **1993**, *97*, 1674.

204. Seydel, U.; Lindner, B. *Fresenius Z. Anal. Chem.* **1981**, *38*, 253.

205. Lindner, B.; Seydel, U. *Int. J. Mass. Spectrom. Ion Phys.* **1983**, *47*, 265.

206. Spey, B.; Bochem, H. *Fresenius Z. Anal. Chem.* **1981**, *38*, 239.

207. Pogue, R.T.; Zhang, W.; Wonkka, R.E.; Majidi, V. *Microchem. J.* **1994**, *50*, 301.

208. Eloy, J.F. *Scanning Electron. Microsc.* **1985**, *II*, 563.

209. Hercules, D.M.; Novak, F.P.; Viswanadhan, S.K.; Wilk, Z.A. *Anal. Chim. Acta* **1987**, *195*, 61.

210. Hutt, K.W.; Housden, J.; Wallach, E.R. *Int. J. Mass Spectrom. Ion Proces.* **1989**, *94*, 237.

211. Hurst, G.S.; Nayfeh, M.H.; Young, J.P. *Appl. Phys. Lett.* **1977**, *30*, 229.

212. Hurst, G.S.; Payne, M.G.; Kramer, S.D. Young, J.P. *Rev. Mod. Phys.* **1979**, *51*, 767.

213. Hurst, G.S.; Payne, M.G. In *Principles and Applications of Resonance Ionization Spectroscopy;* Adam Hilger: Bristol and Philadelphia, 1988; pp. 205–247.

214. Nogar, N.S.; Estler, R.C.; Fearey, B.L.; Miller, C.M.; Downey, S.W. *Nucl. Instr. Meth.* **1990**, *B44*, 459.

215. Sjöström, S.; Mauchien, P. *Spectrochim. Acta Rev.* **1993**, *15*, 153–180.

216. Bakos, J.; Ignacz, P.N.; Kedves, M.A.; Szigeti, J. *Optical Eng.* **1993**, *32*, 2487.

217. Kedves, M.A.; Bakos, J.S.; Földes, I.B.; Ignacz, P.N.; Kocsis, G. *Nucl. Instr. Meth.* **1990**, *B47*, 296.

218. Gray, A.L. *Analyst* **1985**, *110*, 551.

219. Arrowsmith, P. *Anal. Chem.* **1987**, *59*, 1437.

220. Longerich, H.P.; Jackson, S.E.; Fryer, B.J. *Geosci. Can.* **1993**, *20*, 21.

221. Fryer, B.J.; Jackson, S.E.; Longerich, H.P. *Can. Mineral.* **1995**, *33*, 303.

222. Brenner, I.B.; Zhu, J. *Fresenius J. Anal. Chem.*, **1996**, *355*, 774.

223. CETAC Technologies, Inc., 5600 South 42 Street, Omaha, Nebraska, 68107 U.S.A.

224. Cousin, H.; Weber, A.; Magyar, B.; Abell, I.; Günther, D. *Spectrochim. Acta* **1995**, *50B*, 63.

225. Arrowsmith, P. In *Lasers and Mass Spectrometry;* Lubman, D.M., Ed.; Oxford University Press: New York, 1990; Chapter 8.

226. Arrowsmith, P.; Hughes, S.K. *Appl. Spectrosc.* **198**, *42*, 1231.

227. Gießmann, U.; Greb, U. *Fresenius J. Anal. Chem.* **1994**, *350*, 186.

228. Date, A.R.; Gray, A.L. *Application of Induced Coupled Plasma Mass Spectrometry;* Blackie: Glasgow and London; and Chapman & Hall: New York, 1989; Table 1.3.

229. Horlick, G.; Shao, Y. In *Inductively Coupled Plasmas in Analytical Atomic Spectrometry;* Montaser, A.; Golightly, D.W., Eds.; VCH Publishers: New York, 1992; Chapter 12.

230. Hager, J.W. *Anal. Chem.* **1989**, *61*, 1243.

231. Broadhead, M.; Broadhead, R.; Hager, J.W. *Atom. Spectrosc.* **1990**, *11*, 205.

232. Dalpé, C.; Baker, D.R. *Can. Mineral.* **1995**, *33*, 481.

233. Durrant, S.F.; Ward, N.I. *Fresenius J. Anal. Chem.* **1993**, *345*, 512.

234. Federowich, J.S.; Jain, J.C.; Kerrich, R. *Can. Mineral.* **1995**, *33*, 469.

235. Foord, E.E.; da Sa C. Chaves, M.L.; Lichte, F.E. *Mineral. Rec.* **1994**, *25*, 133.

236. Imai, N.; Yamamoto, M. *Microchem. J.* **1994**, *50*, 281.

237. Kontak, D.J.; Jackson, S. *Can. Mineral.* **1995**, *33*, 445.

238. Lichte, F.E. *Anal. Chem.* **1995**, *67*, 2479.

239. Ludden, J.N.; Feng, R.; Gauthier, G.; Stix, J.; Shi, L.; Francis, D.; Machado, N.; Wu, G. *Can. Mineral.* **1995**, *33*, 419.

# 5

# Laser Induced Plasmas for Analytical Atomic Spectroscopy

*Yong-Ill Lee, Kyuseok Song, and Joseph Sneddon*

## 5.1 Introduction

When the output from a (pulsed) laser is focused onto a small spot of a solid surface, an optically induced plasma, frequently called a laser-induced plasma (LIP) or laser-ablated plasma (LAP) or laser spark is formed at this surface. The plasma will be formed when the laser power density exceeds the breakdown threshold value of the solid surface. This plasma has been used for sampling, atomization, excitation, and ionization. In analytical atomic spectroscopy, it has been frequently used and proposed as a source for atomic emission spectrometry. In this case the technique is most often referred to as laser-induced breakdown (emission) spectrometry (LIBS). It has been most commonly proposed and used for the direct determination of elements in solid samples but can be used for liquids and/or solutions and gases.

The advantage of LIBS is in the minimal sample preparation required for a solid sample (resulting in increased throughput and reduction of tedious and time-consuming preparation procedures which can lead to contamination), ability to analyze conducting as well as nonconducting solid samples, ability to analyze extremely hard materials that are difficult to digest or dissolve, such as ceramics and superconductors, potential for multi-element determinations, and potential for direct detection in aerosols (a solid or liquid particle in a gaseous medium) or ambient air using a transient plasma. Disadvantages include increased cost and complexity of the system, difficulty in obtaining suitable standards (for this reason the technique must be regarded as semiquantitative), interference effects (including matrix and, in the case of LIBS in aerosols, the potential interference of particle size), and poorer sensitivity (usually) than several competing atomic spectroscopic techniques using solutions such as inductively coupled plasma atomic emission spectrometry (ICP-AES), inductively coupled plasma mass spectroscopy (ICP-MS), and graphite furnace atomic absorption spectrometry (GFAAS).

This chapter describes the basic principles, and considerations in producing laser-induced plasma, and includes a discussion on the basic principles, instrumentation and analytical detection techniques used in LIBS, and finally the application to a wide variety of samples.

## 5.2. Laser-Induced Plasma

### 5.2.1 General Features

The interaction of high-power laser light with a target or solid sample has been an active topic not only in plasma physics but also in the field of analytical chemistry. The use of the laser-induced plasma has been reviewed by several authors.[1-9] The book edited by Radziemski and Cremers[1] provides a thorough discussion on LIBS including several chapters on the physics of the plasma. A comprehensive review, including over one thousand references, of the more general topic of the interaction of laser radiation with solid materials and its significance in analytical spectrometry has been published by Darke and Tyson.[2] Radziemski[3] has also reviewed analytical applications of LIBS from 1987 to 1994.

Numerous factors affect the ablation process, including the laser pulse properties, such as pulse width, spatial and temporal fluctuations of the pulse, and power fluctuations. The mechanical, physical, and chemical properties of the sample also influence the ablation process. The phenomena of laser-target interaction have been reviewed by Grey-Morgan[10] and Ready.[11] Ready[11] gave a comprehensive description of melting and evaporation at metal surfaces. Anisimov and his coworkers[12,13] related the thermal conductivity mechanism to the boundary condition of free vaporization of the solid into a vacuum. Caruso et al.[14] found different regions existing in a metal and the hot plasma formed on the outer surface expanding toward the light source at a supersonic speed.

The hot vapor plasma interacts with the surrounding atmosphere in two ways: (1) the expansion of the high-pressure vapor drives a shock wave into atmosphere, and (2) energy is transferred to the atmosphere by a combination of thermal conduction, radiative transfer, and heating by the shock wave. The subsequent plasma evolution depends on irradiance, size of vapor plasma bubbles, target vapor composition, ambient gas composition and pressure, and laser wavelength. The history of important quantities such as radiative transfer, surface pressure, plasma velocity, and plasma temperature is strongly influenced by the nature of the plasma, as is the final steady-state nature of the plasma.

The plasma is initiated and sustained by inverse bremsstrahlung absorption during collisions among sampled atoms and ions, electrons, and the gas species. This is the reverse of the well-known bremsstrahlung process in which high-energy electrons, upon traversing a gas or solid, emit radiation as they slow down (from the German words "bremsen," to slow down, and "strahlung," radiation).[15] The theoretical considerations on plasma production and heating by means of laser beams have been proposed by several workers.[16-22] Shah et al.[20] reported a modeling study of species densities in laser ablation from a surface. They used blast wave theory with ionization described by the Saha equation to calculate the time-dependent spatial distribution of the density of several species from a plasma formed on a carbon surface. These models produce essentially similar solutions for the plasma temperature, density, and expansion velocity and are broadly in agreement with experimental results. Cottet and Romain[21] studied the formation and decay of laser-generated shock waves by a hydrodynamic model. Measurements of shock wave velocities were performed on copper foils for incident intensities between $3 \times 10^{11}$ and $3 \times 10^{12}$ W/cm$^2$, with the use of piezoelectric detectors. Balazs et al.[22] calculated the time dependence of density, velocity, temperature, and pressure profiles below and above the plasma ignition threshold. Below the plasma ignition threshold, the temperature of the expanding plume never exceeds the surface temperature, and in the vapor, thermal ionization is almost completely absent. The plume expands into the vacuum, and its flow becomes supersonic. In the high-fluence case, the energy delivered to the plume through electron-neutral inverse Bremsstrahlung processes was enough to elevate the temperature close to the surface value. This gives rise to high electron density as well as intense light absorption.

The temporal history of a laser plasma in a gas or on a solid target is illustrated in Figure 5.1. In liquids the temporal history is compressed. In solids in vacuum, higher stages of ionization are reached for the same intensity on the target. Because of the high initial density of free electrons and ions, the spectral broadening is dominated early on by the Stark effect. As time progresses in a single laser-generated plasma, recombination occurs, the electron density decreases, and pressure broadening (Stark effect due to neutrals) is often the main determiner of linewidth. The pressure and nature of the ambient gas influences the absolute line intensities, the linewidths, and in some cases, relative line intensities due to near-resonant collisions. Given the temporal behavior of the plasma, time resolving the signal in some way is virtually required if analytical information is to be obtained from direct observation. More attention is being paid to the effect of wavelength now that UV laser sources are available.

## 5.2.2 Factors Influencing Plasma Production

### 5.2.2.1 *Laser Parameters*

#### 5.2.2.1.1 Influence of Irradiation Wavelength

The interaction of different laser systems that irradiate from the near- and mid-infrared to the ultraviolet region with solid materials has been studied exten-

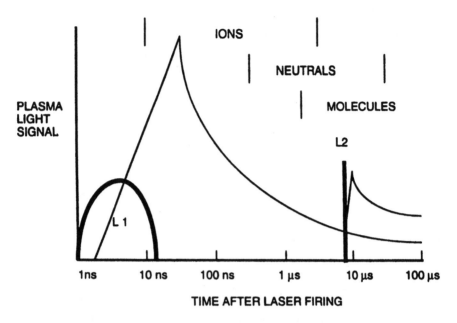

**Figure 5.1** Typical temporal history of a laser-induced plasma (Nd: YAG laser: 5 mJ, $\lambda = 1064$ nm, $10^9$–$10^{11}$ W/cm$^2$), illustrating the temporal regimes where various spectra usually occur. The laser pulse is labeled L1. A second pulse (L2), with variable delay from the first, is also used sometimes. From reference 3.

sively, and the influence of the laser wavelength on the material removal was reported by several authors. Eloy[23] researched laser evaporation with different wavelengths and reported that the thickness of a car paint sample can be reduced many times by the use of a UV wavelength compared to red laser light, to obtain analytical signals of the same magnitude on mass spectrometry. Bingham and Salter[24] researched the ion production in mass spectrometry using three different lasers: $CO_2$, ruby, and Nd:YAG. They obtained the highest sensitivity for the elements P, S, Ti, V, Cr, Mn, Ni, Co, Cu, As, Zr, Mo, Nb, S, Ts, and W, with a steel standard using ruby laser ($\lambda = 694$ nm) ablation. However, the $CO_2$ laser ($\lambda = 1064$ nm) showed poor sensitivity for some high boiling point elements (Ti, V, Zr, Mo, Nb, Ta, and W). Fabbro et al.[25] used a Nd:YAG laser that was frequency doubled and quadrupled to give wavelengths of 1064, 532, and 266 nm to study the effect of wavelength, and postulated the following equation for the mass ablation rate, $m$ (kg/s cm$^2$), its dependence on wavelength, $\lambda$, and the absorbed flux, $F_a$ (W/cm$^2$):

$$m = 110 \, (F_a^{1/3})(10^{14})\lambda^{-4/3}. \tag{5.1}$$

They found that the mass ablation rate increased strongly at shorter wavelengths. Measurements of the ablation pressure generated by the ablating plasma have been carried out at a number of laboratories.[26–28] These results confirmed the expected higher ablation pressure with shorter wavelength laser irradiation.

Kwok et al.[29] investigated the optical emission produced by laser ablation of $YBa_2Cu_3O_7$ targets using a wide range of laser wavelengths and showed that 193-nm radiation produced mostly neutral atomic species while 1064- and 532-nm radiation produced mostly ionic species. Recently, Kagawa and his co-workers[30,31] used two different lasers, an XeCl excimer laser and a TEA-$CO_2$ laser, to produce laser-induced plasma on the surface of glass samples. The LIP could not be generated by the TEA-$CO_2$ laser because of the lack of expulsion from the sample, however, with the use of the excimer laser, LIP was generated and, thus, it is amenable for atomic spectroscopic analysis.

### 5.2.2.1.2 Influence of Irradiation Energy

In order to describe the fate of the laser energy during laser-solid interactions, several processes should be considered. Due to the character of the target material, a fraction of the energy is absorbed from the laser pulse while the rest is reflected by the surface. The deposited part of the laser energy is converted into local heat instantaneously, which can in turn diffuse by heat conduction. An increase in temperature may induce appreciable changes in optical and thermal properties of the target material, thus influencing the rate of energy deposition and heat transfer. If the surface temperature is sufficiently high, a phase change (melting) may occur and part of the absorbed laser power will be expanded into the latent heat of transition.

For most materials, the power density required for evaporation is in the range of $10^4$–$10^9$ W/cm². In the range of $10^4$–$10^7$ W/cm², the resulting vapor consists of polyatomic particles.[32] Selter and Kunze[33] studied the degree of atomization in laser-produced vapor from titanium targets. At power densities below $7 \times 10^7$ W/cm², no atoms were observed in the vapor, whereas the evaporated material became partially ionized above $5 \times 10^8$ W/cm². It was assumed that a power density in the range of $10^6$–$10^8$ W/cm², depending on the solid target, was sufficient for analytical measurements in a laser-ablated plume.[34] Carroll and Kennedy[35] also found that the threshold power density for the formation of a plasma plume by the laser irradiation was typically in the neighborhood of $10^8$ W/cm². The threshold power density varies with the wavelength of a laser primarily because the absorbance of the target surface depends on the wavelength of the incident light. Dyer[36] determined the threshold energy for the generation of a plasma on a copper target of $10^8$ W/cm² with a KrF-excimer laser ($\lambda = 248$ nm).

The criterion for the plasma ignition threshold as a relation of the plasma absorption coefficient by the adiabatic absorption model for $CO_2$, ruby, and quadrupled Nd:YAG lasers and for different materials was established by Vertes et al.[37] They found an increase of the threshold for materials with increasing ionization potential. It was also noticed that in similar circumstances the threshold temperature for the frequency-quadrupled Nd:YAG ($\lambda = 266$ nm) laser was always the largest, while for the $CO_2$ laser it was the lowest. This is in accordance with the widely known observation that UV lasers produce sharply etched craters in the target, while increasing laser wavelength creates a molten crater rim.

Sneddon and coworkers[38-41] researched quantitative plasma emission with increasing laser energy by an ArF-excimer laser with the selected metals, Zn, Cu, Ni-alloy, and Fe-alloy. The results show that the emission intensity from a laser-induced plasma with increasing laser energy can be quantified both theoretically and experimentally, and a potential correlation obtained. Based on a heat conduction mechanism, they derived the following equation:

$$m(t) = A \, (aIt) + B \, (aI)^2 \, t^{3/2}. \tag{5.2}$$

This equation shows that the mass removal, $m(t)$, is proportional to the metal thermal properties $(A,B)$, energy coupling factor $(a)$, laser irradiance $(I)$, and laser pulse duration $(t)$.

Iida and Yeung[42] studied the ablation mechanism by the optical monitoring of the LIP derived from graphite using a Q-switched Ng:YAG laser at various power densities (0.6–1.8 GW/cm²) and proposed an ablation mechanism of graphite in a dissociative model shown in Figure 5.2. The carbon atoms and small molecules are omitted in the figure for simplicity, but they are formed in the ablation stage and also after ablation by dissociation from the carbon particles. At low power density (Figure 5.2a), relatively large particles come out from the graphite target surface by interlayer dissociation. On the other hand, at higher power densities (Figure 5.2b), the particles are decomposed into smaller clusters with increased probability of intralayer bond breakage. The addition of a helium atmosphere (Figure 5.2c) increases the temperature and density of the laser-induced plasma and promotes decomposition. Thus, carbon particles of smaller sizes will be produced.

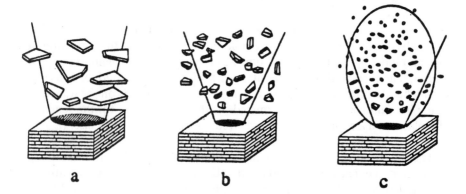

**Figure 5.2**   Schematic representation of laser ablation of a graphite target and thermal modification of the ablated particle. **(a)** At low power densities, **(b)** at high power densities, **(c)** at high power densities under helium atmosphere. From reference 42.

### 5.2.2.2 *Physical Properties of the Target Material*

The physical properties of the target materials have an important influence on the shape and size of craters in target materials. The reflection of part of the laser energy is an important consideration in determining the fraction of laser energy absorbed by sample materials.

Allemand[43] showed that the reflectivity of the sample surface, density, specific heat, and boiling point of the pure metal target have an important influence on the size of the craters, and derived the relationship to the physical constants of pure materials.

$$D = A (1 - R)/\rho C_p T_b \qquad (5.3)$$

where $D$ = diameter of total splash (crater), $A$ = proportionally constant (in energy per unit area), $R$ = reflectivity of the surface at 1 mm, $\rho$ = density, $C_p$ = specific heat, $T_b$ = boiling temperature.

Ishizuka[44] studied the size and depth of the crater in samples of rare earth oxides, aluminum oxide, and sodium salts by using a Q-switched ruby laser. The crater produced by a laser shot was about 1 mm in diameter regardless of the composition of the matrix, but the depth of the crater depended on the type of matrix. A comparison of the crater size of the homogeneous material revealed that the thermal conductivity is an important parameter. The depth of the craters increased with this value. The volume heated depended on the thermal conductivity of the material under the same laser conditions.[45] On the other hand, heating of material around the crater increased the incident light intensity because evaporation depends only on the boiling point of the material at a fixed pressure.

Dimitrov et al.[46,47] investigated the substance evaporation processes and the kinetics of plasma plume development depending on target orientation with respect to the laser radiation source direction. When the metal target is irradiated by laser radiation, the erosion products emerge nearly perpendicular to the target surface. When the target surface is inclined with respect to the direction of laser radiation, the path length of the radiation in the plasma is shortened, which results in decreased absorption of the laser-induced plasma.

Lee et al.[48] looked at the ablation caused by an ArF excimer laser on copper and lead. The irradiance was on the order of $1.5 \times 10^9$ W/cm². They concluded that copper, which has high thermal conductivity and a high boiling point, produced a more confined plasma with a high excitation temperature. While lead, with a low thermal conductivity and boiling point, generated an extended plasma with a low excitation. They continued their study[49] of the relationship between the argon gas breakdown and the physical properties of target materials by using five metals, Cu, Zn, Fe, Al, and Pb, as a diagnostic tool. They postulated the following relationship:

$$I_B = K \frac{1}{C_p C_T \rho} \qquad (5.4)$$

Where, $I_B$ = intensity of argon gas breakdown emission (counts), $K$ = proportionality constant, $C_p$ = specific heat (J/K mol), $C_T$ = thermal conductivity (W/m K), and $\rho$ = density (kg/m³).

### 5.2.2.3 Ambient Conditions

The atmospheric influences on the LIP were concerned with the mass loss, crater formation, and plasma emission characteristics. Several workers have studied the effect of atmospheric conditions on laser ablation atomic emission spectrometry (LA-AES). The researches were mostly carried out in air, argon, helium, and nitrogen. Iida[50] researched the emission of the laser-induced plasma, with the use of a Q-switched ruby laser of energy of 1.5 J in a 20-ns duration in an argon atmosphere at reduced pressure. The emission intensities of atomic lines increased several fold in an argon atmosphere, in comparison with those obtained in air at the same pressure. Moderate confinement of plasma and a resultant increase of emission intensities were achieved at 50 Torr. They also used a Q-switched Nd:YAG laser (150 mJ/pulse, 10-ns pulse) to study the effect of atmosphere and power density on plasma generation.[51] They found that tight focusing of laser radiation did not directly bring about a plasma of high emission intensity, because of the absorption of laser energy by the plasma itself. The importance of prevention of a gas breakdown before sample vaporization was also indicated. Grant and Paul[52] studied the laser-induced plasma by irradiation of a steel target with an XeCl excimer laser ($\lambda$ = 308 nm) with an energy of 40 mJ/pulse in an atmosphere of air, argon, nitrogen, and helium at pressure from 0.5 to 760 Torr. The maximum spectral intensity and line-to-background (L/B) ratio occur in an atmosphere of argon at a pressure of 50 Torr.

Lee et al.[53] investigated the effect of pressure over the range of 10–760 Torr and of atmosphere (air, argon, and helium) on an ArF excimer laser– ($\lambda$ = 193 nm, 100 mJ/pulse) induced plasma created above the surface of a copper target. With the use of neutral copper lines, reduced pressure (from 760 to 10 Torr) resulted in a 7-fold increase in air and an 11-fold increase in an argon atmosphere. The use of helium resulted in only a 1.5-fold increase over that obtained at 760 Torr. They also observed nitrogen in air and argon breakdown in the plasma, and concluded gas breakdown influences the laser energy coupling to the metal target. However, helium gas breakdown was not observed because of higher ionization potential and high thermal conductivity compared to argon and nitrogen. This characteristic peak of neutral copper atom lines is shown in Figure 5.3. The differences in shape and appearance of the sequence of the plasma images collected by a CCD imaging detector were dependent on argon and helium at atmospheric pressure, illustrated in Figure 5.4.

Kagawa et al.[55] studied an XeCl excimer laser– (15–70 mJ/pulse, 20 ns pulse duration) induced shock wave plasma on a Zn plate in a surrounding gas at low pressures (0.75–11.3 Torr) and defined the role of surrounding gas as only a damping material to prevent the free expansion of the propelled atoms. The total emission intensity of the atom emission lines is determined mainly by the

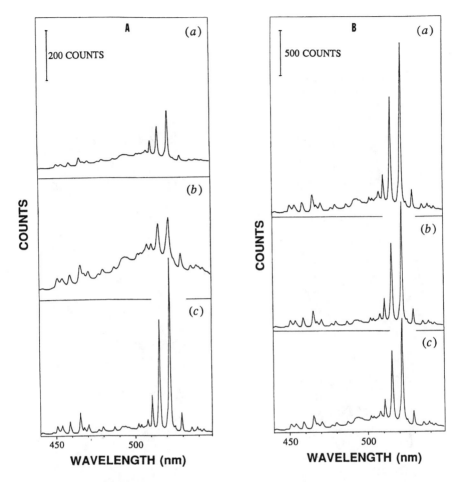

**Figure 5.3**   Emission spectra of a laser-induced plasma from the center of the plasma 760 Torr, **(A)**, and 50 Torr, **(B)**, and under various atmospheres: argon, **(a)**, air, **(b)**, and helium, **(c)**. From reference 53.

amount of propelling atoms and the entire amount of kinetic energy they produce. Mao et al.[56] also demonstrated the shielding effect on the coupling of laser energy to a target surface during the interaction of picosecond pulsed laser beam and target material using a He and Ar gas atmosphere. They concluded that Ar is easier to ionize than He, and plasma shielding is more severe in Ar than in He. Kuzuya et al.[57] studied the effect of laser energy and atmosphere on the emission characteristics of LIP with the use of a Q-switched Nd:YAG laser over a laser energy range of 20–95 mJ/pulse. The experimental results showed that the maximum spectral intensity was obtained in argon at around 200 Torr at high laser energy of 95 mJ/pulse, whereas the line-to-background (L/B) ratio was maximized in helium at around 40 Torr at a low energy of 20 mJ/pulse.

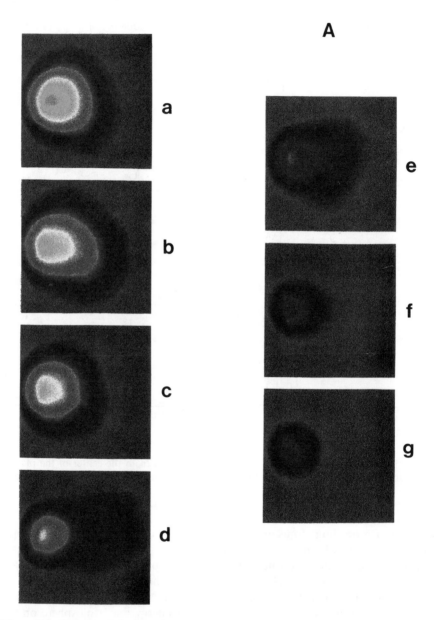

**Figure 5.4** Images of the laser-induced plasma formed with copper under helium (**A**) and argon (**B**), and at different pressures: 760 Torr (**a**), 600 Torr (**b**), 400 Torr (**c**), 200 Torr (**d**), 100 Torr (**e**), 50 Torr (**f**), and 10 Torr (**g**). From reference 54.

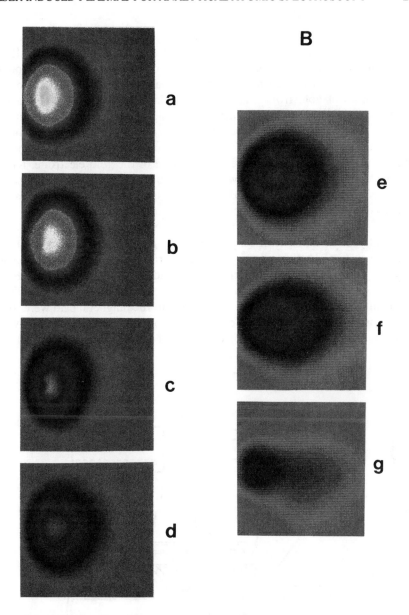

**Figure 5.4** (*Continued*).

## 5.2.3 Influence of Electric and Magnetic Fields on Plasma Emission

Over the past decade, there has been interest in the use of electric or magnetic fields for enhancing the analytical characteristics of plasma sources for atomic spectroscopic analysis. Pulsed magnetic fields have been used to alter the properties of microwave plasmas,[58] spark discharges,[59,60] and exploding conductor plasmas.[61–66] A few recent studies have looked at the effect of a static electric field on ultraviolet emission enhancement and breakdown threshold. Hontzopoulos et al.[67] used a 3–20 mJ KrF excimer laser to study the effect of UV emission lines of gold from a laser-induced plasma on the surface of a gold target in fields up to 13 kV/cm. They found a significant enhancement, up to a factor of 100, for some lines above 6.6 kV/cm, and saturation for some lines at 20 kV/cm. They tentatively interpreted this enhancement on the basis of recombination processes taking place near the surface of the gold electrode. Kumar and Thareja[68] studied the breakdown in Ne, Ar, and Xe gas at different pressures using a XeCl excimer laser (60 mJ, 8-ns pulse length) up to a maximum field strength of 1000 V/cm. They concluded that their results were similar to those observed with a high-power laser alone.

Recently, Mason and Goldberg[69] designed and constructed a new capacitive discharge system. The pulsed magnetic field, produced by capacitive electrical discharge through a specially designed solenoid, was oriented normal to the laser axis. The schematic diagram of this system is shown in Figure 5.5. Temporally integrated emission enhancements due to the magnetic field were found to be most significant when the plasma was formed about 1 mm below the magnetic field axis. The degree of confinement of the plasma increased with

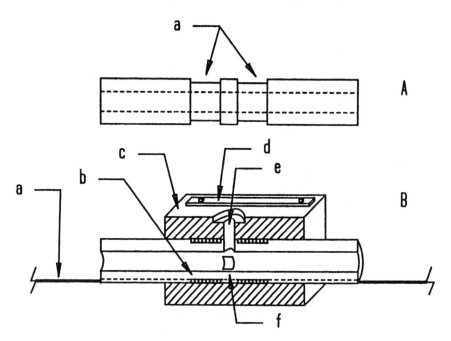

**Figure 5.5** **(A)** Coil support tube: **(a)** recessed groove for the two halves of the coil; **(B)** Coil: **(a)** write for connection to capacitor bank, **(b)** coil, **(c)** polycarbonate coil support blocks, **(d),** steel clamp support, **(e),** laser access port; **(f)** side-on viewing port. From reference 69.

magnetic field strength. There was also an increase in line broadening, neutral atom self-reversal, and minor constituent emission intensities.[70]

They subsequently conducted research[71] on the dynamic effects of a high-intensity pulsed field having a maximum strength of 85 kG, by using time-resolved emission and absorption measurements. Spatial and temporal discrimination of emission enhancement indicated that radial compression was due to static magnetic field interactions with the laser-induced plasma and that mild Joule heating from the small induced current was most likely responsible for emission enhancements later in time. They concluded that more efficient coupling of energy from the magnetic field to the plasma would require low-pressure operation in a controlled atmosphere and/or a pulsed magnetic field having a greater $dB/dt$ ($B$ is the magnetic field intensity). The influence of a magnetic field on the plasma emission of Mg induced by a KrF excimer laser was reported by Dirnberger et al.[72] In certain spatial regimes, they observed a dramatic enhancement of emission signal from neutral and ionized species over the field-free case.

### 5.2.4 Excitation Temperatures and Electron Densities of the Plasma

Temperature is one of the more important properties of any excitation source. Knowledge of the temperature of an excitation source is vital to the understanding of the dissociation, atomization, ionization, and excitation processes occurring in the source and is helpful in attempts to utilize the source to its maximum analytical potential. The methods most frequently used for determination of excitation temperatures are the two-line method[73–75] and the Boltzmann plot method.[48,76–78] Following the Boltzmann distribution and neglecting self-absorption, the line intensity is related to $T_{exc}$ by Equation 5.5:

$$I_{kl} = \left(\frac{hc}{\lambda}\right)(n_o)\left(\frac{g_k}{g_o}\right)(A_{kl})\exp\frac{-E_k}{kT_{ex}} \tag{5.5}$$

where $l_{kl}$ refers to the line intensity for the transition between the upper level $k$ to the lower level $l$ of the emitting species, $T_{exc}$ is the excitation temperature, $g_k$ and $g_o$ are the statistical weight factors for the excited and the ground states, respectively, $n_o$ is the population of atoms in the ground state. $A_{kl}$ stands for the probability of the transition per unit of time (Einstein transition probability), $E_k$ is the energy of the excited state, $h$ ($6.626076 \times 10^{-34}$ J s) is Planck's constant, $k$ ($1.38066 \times 10^{-23}$ J/K) is the Boltzmann constant, and $C$ ($2.9979 \times 10^8$ m/s) is the speed of light in a vacuum. In the Boltzmann plot method, if relative intensity for a given species is measured at various wavelengths, Equation 5.5 can be expressed as the following:

$$\ln\left(\frac{I_{kl}\lambda}{g_k A_{kl}}\right) = -\frac{E_k}{kT}. \tag{5.6}$$

Therefore, a plot of the In ($I\lambda/gA$) versus $E_k$ gives a means of measuring the electronic excitation temperature. In the two-line method, the ratio of the intensities of two spectral transitions for given species is provided by the following equation:

$$\frac{I_l}{I_k} = \left(\frac{g_l}{g_k}\right)\left(\frac{A_l}{A_k}\right)\left(\frac{\lambda_k}{\lambda_l}\right) \exp[(E_k - E_l)/kT. \tag{5.7}$$

Grant and Paul[52] determined the electron temperature and density of XeCl excimer laser–induced plasma. The relative atomic contents of the eleven Fe(1) lines used in the Boltzmann plot of electron temperature were used. The temperature ranged from 9000 to 22,000 K depending on the ambient conditions. Temperature decreased with distance from the surface and with decreasing ambient pressure. Electron densities were calculated according to the Saha equation and the fitted values of temperature with assumption of local thermodynamic equilibrium (LTE). The density profile exhibited features similar to those for temperature, ranging from $3 \times 10^{19}$ to about $10^{16}$ cm$^{-3}$. Kagawa et al.[79] calculated the excitation temperature in a high-power nitrogen laser–induced plasma with the two-line method by the line pair of Zn(I). The temperature ranged from 8000 to 9000 K, and the region of maximum was at a point some distance from the center of the plasma rather than at the center. Ursu et al.[80] studied the optical breakdown of plasma in a gas in front of various solid samples by a TEA-CO$_2$ laser source. They measured the energy absorbed into a blade calorimeter placed at various distances from the center of the plasma. An upper limit for the vapor temperature of ~14,000 K was inferred from the characteristic darkening curve on the photographic film. They found that the initial maximum corresponding to the breakdown plasma in gas having a temperature of ~20,000 K is followed by a luminescence tail due to vapors acting with the fireball and having a temperature of ~10,000 K.

Radziemski and coworkers[81,82] measured the temporal variation of temperature and electron density in an air plasma induced by a CO$_2$ laser operating at 0.5 and 0.8 J/pulse. The excitation temperature was determined by Boltzmann plot, and ranged from 19,000 K at 1 μs to above 11,000 K at 25 μs. The electron density was measured at 500 mJ/pulse and determined to be $3.6 \times 10^{17}$ cm$^{-3}$ at 1 μs and $4 \times 10^{16}$ cm$^{-3}$ at 25 μs. Lee et al.[48,53] calculated the excitation temperature of an ArF excimer laser–(100 mJ/pulse, 11-ns pulse length) induced plasma by the Bolzmann plot method with Cu(I) and Pb(I) lines. Typical Bolzmann plots for the copper and lead plasma are shown in Figure 5.6. The temperatures of the excimer laser–induced plasma were quite high, ranging from 13,200 to 17,200 K in the plasma formed with copper and from 11,700 to 15,300 K for the plasma formed with lead, depending on the location in the plasma. They also measured the excitation temperatures of the plasma under different pressure and atmosphere. The temperatures ranged from about 14,000 K at 10 Torr to 18,000 K at 760 Torr for an air atmosphere, from about 13,400 K at 10 Torr

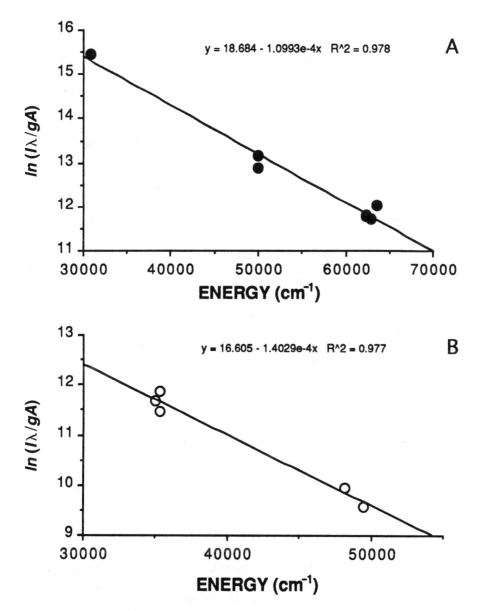

**Figure 5.6** Typical Bolzmann plot [ln (I$\lambda$/gA) vs. energy] for **(A)** Cu(I) spectral lines and **(B)** Pb(I) spectral lines. From reference 48.

to 14,200 K at 760 Torr for an argon atmosphere, and from 12,600 K at 760 Torr to 14,800 K at 200 Torr for a helium atmosphere.

Measurements of precision and accuracy in the plasma temperature are highly dependent on sample composition, homogeneity, surface condition, and particle size. Precision is typically 5–20% but values of less than 1% have been achieved under certain conditions.[83] However, the accuracy of the estimation calculated with the Bolzmann equation is uncertain. It depends on the existence of the local thermodynamic equilibrium (LTE) assumed for this method and the accuracy of spectroscopic constants of neutral atom lines.

### 5.2.5  Use of Laser-Induced Plasma as a Continuum Source for Atomic Spectroscopy

Because the continuum emission of the LIP is very intense, the use of this plasma continuum source for absorption spectroscopy in the visible and near UV region was proposed by Evtushenki and Ostrovskaya.[84] They evaluated the plasma intensity in helium and argon, in the range of 300–700 nm and found that the plasma formed in argon was the most intense. An increase in plasma intensity with an increase in gas pressure was also reported. Furthermore, they utilized the plasma continuum source to study the vaporization of $CaCl_2$ by absorption spectroscopy. Adamson and Cimolino[85] utilized LIP for an infrared absorption study of bond stretching within transition metal complexes. This research demonstrated that the laser plasmas were at least 25 times more intense than a typical glow bar used in conventional infrared absorption spectroscopy. They also observed that the plasma emission lifetime, when formed in air, was 150 ns in the visible region and 2 μs in the infrared region. In argon, the plasma emission lifetime was 4 ms in the infrared region.

Laporte et al.[86] used the LIP in rare gases as a vacuum UV continuum source (110–170 nm). Working with the Nd:YAG laser and rare gases at pressures of $1.00 \times 10^5$ to $6.00 \times 10^5$ Pa, they reported photon densities of $5 \times 10^8$ photon $nm^{-1}s^{-1}$ for argon and $10^9$ photon $nm^{-1}s^{-1}$ for krypton, near 170 nm. The continuum emission duration lasted from 20 ns for neon to 40 ns for krypton. The utilization of LIP as a continuum source in the soft X-ray region was also reported by Carroll et al.[87]

Recently, Majidi et al.[88] illustrated the utility of LIP for transient molecular absorption studies in electrothermal atomizers. These studies demonstrated the feasibility of laser plasma as a continuum source in the UV region. Their research[89] investigated the LIP as a continuum source by time- and wavelength-resolved LIP emission spectra in helium, argon, nitrogen, air, and a helium/argon mixture in the spectral range of 200–700 nm. A radiant intensity of $8 \times 10^{11}$ photons $pulse^{-1}$ $s^{-1}$ was produced at 633 nm with 240 mJ/pulse from a Nd:YAG laser. Fundamental characterization of LIP in gases by spectroscopic means continues. Simeonsson and Miziolek[90] reported on spectroscopic characterization of microplasma formed by ArF laser pulses. The target gases were CO, $CO_2$, methanol, and chloroform. They speculated that the absorption process in gas streams containing atomic carbon would have very low breakdown thresholds, from microjoules to millijoules.

## 5.3 Laser-Induced Breakdown Spectrometry

### 5.3.1 Basic Principles

Elemental analysis based on the emission from a plasma generated by focusing a powerful laser beam on a sample (solid, liquid, or gas) is known as laser-induced breakdown spectrometry (LIBS). The principles of LIBS are similar to those of conventional plasma atomic emission spectrometry, such as inductively coupled plasma (ICP)-AES, microwave-induced plasma (MIP)-AES, direct current plasma (DCP)-AES, arc-AES and spark-AES. In atomic emission spectrometry, the light from an excited sample is spectrally resolved and sometimes temporally resolved, as in the case of pulsed light sources, to yield qualitative and quantitative information about the elemental constituents. This distinguishes the LIBS technique from conventional plasma-AES. The sample need not be transported to the plasma source; rather, the plasma is formed in or on the sample in situ. In addition, the LIBS technique can be used to analyze gases, liquids, and solids directly, without sample preparation, because the plasma is produced by optical radiation.

Atomic ions as well as neutral atoms will be produced in plasma sources having high temperatures. Direct observation of the laser-induced plasma with atomic emission spectroscopy is a simple method for analysis of the sample and offers the potential of simultaneous multi-element analysis. The energy levels are unique for each element, so it is possible to perform simultaneous qualitative analysis by examining wavelengths in the emission spectrum of an unknown. For the quantitative analysis, a measurement involves integration of the analytical signals from a line or set of lines of the elements of interest over a certain period of time. The amount of an element present is determined by constructing a calibration curve of signal versus concentration by the use of standard reference materials. Over 1,000 different standard reference materials (SRMs) can be obtained from the National Institute of Standards and Technology (NIST). These materials are certified for their chemical composition, chemical properties, or physical properties by various reliable analytical methods. However, the standard references are often not available with a particular matrix.

### 5.3.2 Instrumentation

#### 5.3.2.1 Lasers

The laser used to produce LIBS must generate pulses of sufficient power to produce the plasma. The most widely used are solid-state lasers such as the Nd:YAG ($\lambda$ = 1060 nm, 532 nm [second harmonics], and pulse length of 5–10 ns) and ruby ($\lambda$ = 693 nm and pulse length of 20 ns) lasers, and gas lasers such as the $CO_2$ ($\lambda$ = 1060 nm and pulse length of 100 ns), $N_2$ ($\lambda$ = 337 nm, pulse length of 30 ps to 10 ns), and excimer ($\lambda$ = 193 nm [ArF], $\lambda$ = 248 nm [KrF], $\lambda$ = 308 nm [XeCl], and pulse length 10–20 ns) lasers. For any of these laser types, it is typical that many tens of millijoules can be produced in time frames of a few nanoseconds. As a result, peak power output is measured in the range from 1 mW to 100 mW. Focusing the laser pulse to a few tens of micrometers in diameter with a simple lens can produce fluxes on the order of $10^{10}$–$10^{12}$ W/cm². The basic principles, properties, and characteristics of the various types of lasers are described in detail in Chapter 2.

The Nd:YAG is the most widely used system due to its reliability and ability to produce powerful laser pulses, particularly when it is Q-switched. The laser pulse energies required to form a plasma depend on many factors including properties of the pulse (such as the energy, mode quality, wavelength, and pulse length) and the material (i.e., reflectivity at the laser wavelength for solids, and density for gases). Typically, 100 mJ/pulse from a Q-switched laser is sufficient to generate the plasma for most types of analysis. Recently, there has been an increase in the use of the excimer laser because of the low UV reflectivity for most metals, resulting in improved energy coupling efficiency, and the high optical resolution offered by the relatively short wavelength.

On the basis of the above work, lasers that irradiate the ultraviolet in wavelength are much better than those that irradiate in the infrared wavelength, and also, a short time of pulse duration is better for the ablation of materials. Following this trend, lasers that irradiate in the X-ray region with femtosecond pulse duration may be even better.

### 5.3.2.2 Sample Container

The sample container is used as a means of reproducibly positioning a sample. A schematic of a typical sample container is shown in Figure 5.7. It is described in detail elsewhere.[38,48]

The laboratory-made quartz chamber is constructed with a gas inlet and gas outlet which allowed an inert gas to be introduced to the sample chamber if desired. This chamber can be easily manipulated and allows convenient changing of the sample materials. The dimensions of the chamber are 60 × 45 × 55 mm; the two long ends are connected to tubes that are one inch in diameter with a length of 1.5 inches. A suitable focusing lens ($f$ = 90 mm) is positioned at one end, and the other end contains the target holder. The target holder is constructed of aluminum, with the cylinder of the holder having two O-rings to ensure a tight fit at the end of the sample holder. In the center of the cylinder, a threaded holder allows the adjustment of the distance between the target and the lens. This is the forerunner of the T-type stainless steel system shown in Figure 5.8 and described in detail elsewhere.[53] Other workers have used various designs.[10–14]

The laser ablation chamber is constructed from 1.5-inch T-type stainless steel

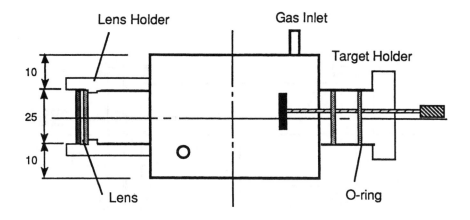

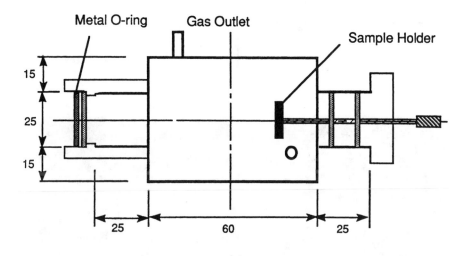

(UNIT: mm)

**Figure 5.7**   Schematic view of the quartz sample chamber system: a plane figure *(top)*, a face figure **(bottom)**. From reference 91.

(Varian NW35CFF) and contains three conflate flange connections. The vacuum seal is made by compressing the copper gasket between the conflate flanges. This is done by tightening six bolts until the first metal-to-metal contact is made between the conflate flanges. This chamber has been used in experiments under vacuum conditions for precise control of ambient gas pressure. A focusing lens is positioned at one end, and the other end contains the target holder. In addition to the lens-containing window, a second window at 90° to the laser beam is used to monitor the plasma emission when the laser beam interacts with the target surface. The window is large enough (diameter: 1.5 inches) that emission

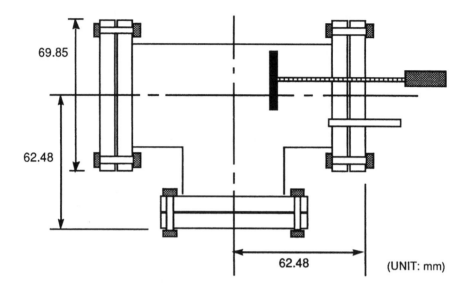

**Figure 5.8** Schematic view of a plane figure *(top)*, and a face-on photograph *(bottom)* of the laser ablation T-type stainless chamber. From reference 91.

from the plasma is not obstructed. In the viewing port, the plane of the quartz window is below the heads of the flange connecting bolts. This design provides wide viewing angles and large, clear diameters, and virtually eliminates the chance of physical damage.

The ablation chamber can be evacuated using a roughing pump, and various ambient gases can be introduced into the chamber. The target holder is constructed of an aluminum rod connected with a rotatable blank flange to ensure a tight fit with the T-type flange. The target holder can be rotated by a stepped motor for ablation of a new surface of the sample each time.

### 5.3.2.3 Detection

The simplest experiment one can do with a laser-induced plasma is to photograph it. The exposure time will be determined by the lifetime of the plasma itself, and the picture is obtained by simply holding the camera shutter open a darkened laboratory while the laser is fired at the sample target. A typical result obtained from an alloy steel target irradiated by a 100-mJ, 11-ns pulse from an ArF excimer laser is shown in Figure 5.9.

Direct photography provides only information of physical interest and, in particular, gives no information about the evolution of the plasma in time. Such information, however, can be obtained by the use of either streak or framing cameras. Various forms of these cameras are available; one in which the image is translated by means of a rotating mirror is shown schematically in Figure 5.10. In a streak camera, a slit is introduced into the optical system so that only a linear element of the plasma can be focused onto the photographic film.

The spectrally resolved signal can be detected with a photomultiplier tube (PMT), photodiode array (PDA), or the recently developed charge transfer device in the form of either a charge-coupled device (CCD) or charge injection device (CID). Early studies of LIBS conducted by Radziemski et al.[93,94] used a variety of photomultiplier tubes with extended ultraviolet or red responses. The same authors used a time-gated linear diode array coupled to a multichannel analyzer (MCA). The array consisted of 1024 diodes in 2.54 cm. Each channel of the MCA recorded the signal seen by one diode during each shot. If not gated, the diodes integrated all the incident energy. The light impinging on the array went through the microchannel plate image intensifier, which had a maximum gain of 2500. With this combination, the array was time-gated by simply switching on the high voltage of the intensifier during the time of interest, typically 200 to several microseconds. The system was sensitive to wavelengths between 350 and 800 nm.

### 5.3.2.4 Analytical Spectrometric Techniques

With various laser systems, the analytical performance, such as accuracy, precision, and detection limit (DL) for direct quantitative atomic spectroscopic analysis are not satisfactory. The primary reason is that the violent, nonlinear

**Figure 5.9** Photograph of laser plume formed at the surface of a metal target with the use of a laser beam with a wavelength of 193 nm and power of 200 mJ/pulse. From reference 38.

laser beam–target material interaction cannot be predicted accurately. The mechanical, physical, and chemical properties of the sample influence the interaction, underscoring a strong matrix effect. The other limitations of LIBS for practical application are self-absorption, line broadening, and the high intensity of the background continuum.[95] These limitations can be minimized by working in a controlled atmosphere and using time-resolved spectroscopic measurement or time-integrated and spatially resolved measurement techniques.

### 5.3.2.4.1 Time-Resolved Spectroscopy

Because pulsed lasers are used in LIBS, the excitation characteristics are a function of time. Temporal history of the plasma can be obtained by recording the

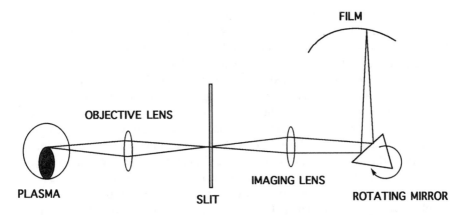

**Figure 5.10**  Principles of the streak camera. Plasma with axis parallel to defining slit.

emission features at predetermined delay times for the gatable intensifier with a variable delay generator that is triggered by the laser pulse. This is easily demonstrated by recording the plasma emission spectrum during distinct intervals as the plasma decays. A typical instrumentation setup used for LIBS is shown in Figure 5.11.

The major components are the laser, a method of focusing the laser pulse onto the sample, a sample container or holder, a system to collect the light from the produced plasma, a method of spectral resolution, a detection system, a method for time resolution, and data storage for the read-out.

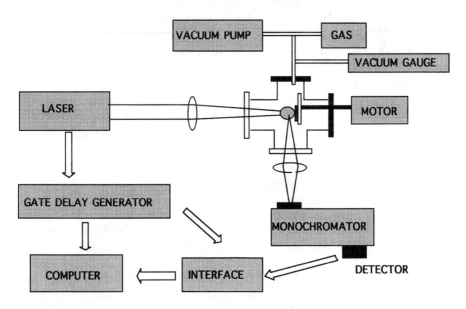

**Figure 5.11**  Instrumental setup for time-resolved LIBS.

Time-resolved spectra of a LIP produced by a Q-switched Nd:YAG laser ($\lambda$ = 1064 nm, up to 250 mJ/pulse, 8-ns pulse duration) in argon atmosphere were recorded by Leis et al.[96] They found that the spectra were significantly dependent on the observation time after the laser pulse. At 500 ns after the laser irradiation, the observed spectrum consisted of a continuum and ion emission lines; at 10 µs, the ion lines had decreased in intensity and the atom lines were more intense. Time-resolved spectrometry of the LIP was also extensively studied by Kagawa et al. by the use of a nitrogen laser ($\lambda$ = 337.1 nm),[97,98] an XeCl excimer laser,[55] 585-nm dye laser,[99] and a TEA $CO_2$ laser.[100] They found an intense continuum emission, called the primary plasma, for a short time just above the surface of the target, and another plasma, called the *secondary plasma,* that expanded with time around the primary plasma, emitting sharp atomic line spectra with a negligible, low background signal. They used the shock wave model to explain the mechanism forming the secondary plasma. However, the appearance of the plasma generated by a Nd:YAG laser is quite different from those in Kagawa's experimental results.[101] The secondary plasma was not observed in the plasma formed under these experimental conditions.

Using time-resolved emission studies, Joseph et al.[102] found that the characteristics of the LIP vary greatly with the composition of the analyte in the focal volume region. The plasma size, stability, and emission properties change with any alteration of the pressure or concentration of phase of the plasma medium. They concluded that helium has define advantages over argon as a buffer gas to support a LIP on a copper or aluminum surface. They used the Nd:YAG laser ($\lambda$ = 1064 nm, 240 mJ/pulse, 7-ns duration) for this research.

The good experimental results obtained by time-resolved plasma emission spectra in argon and helium atmospheres under 760 Torr of gas pressure with copper as a target material are shown in Figure 5.12.[54] The plasma emission was obtained with 1 atm of ambient gas pressure and displayed as a function of wavelength as well as the gate delay time from the trigger of laser system. For this study, an Nd:YAG laser (20 mJ/pulse, $\lambda$ = 532 nm, 8-ns pulse duration) was used for ablating the copper material, and an intensified charge coupled device (ICCD) (1024 $\times$ 256 elements) constituted the detection system. The gate width was 50 ns and the delay time of each spectra was 100 ns.

The sharp analytical atomic lines of copper in the LIP appeared at a delay time of 200 ns under helium atmosphere, and at 2 ms under argon atmosphere. Since the argon atom is a lot bigger than the helium atom, the probability of collision is also larger for argon than helium. The longer lifetime of the laser-induced copper plasma under argon atmosphere relative to that under helium atmosphere can also be understood as a damping effect of particles to prevent plasma diffusion and as a cooling effect by thermal conduction of ambient gas. Therefore, both the damping effect and the cooling effect of buffer gas influence the generation and propagation of the LIP. Figure 5.13 shows the good comparison of copper emission in the LIP depending on the gate delay time under a 1-atm helium atmosphere. The emission peaks show a relatively broad shape

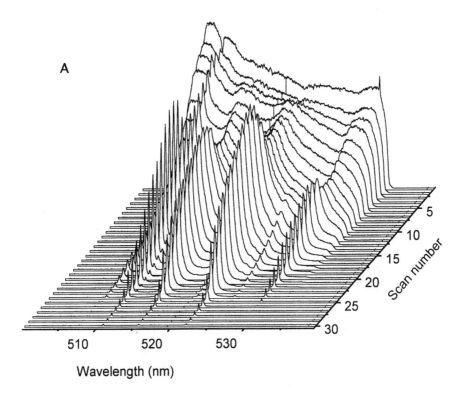

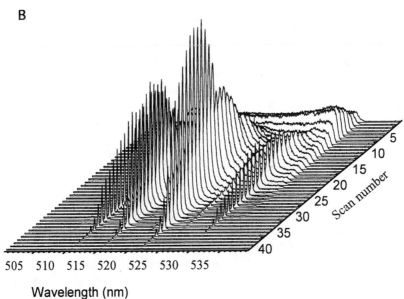

**Figure 5.12**   Time-resolved copper emission spectra from a laser-induced plasma. The plasma spectra were taken from the segment of the plasma located at 0.4 mm from the target surface under (**a**) argon and (**b**) helium atmospheres at 760 Torr.

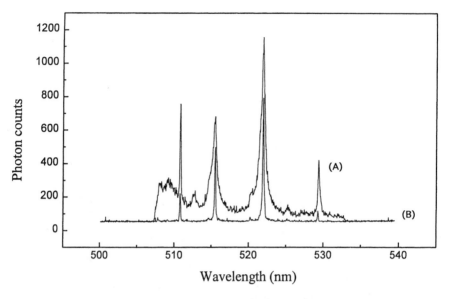

**Figure 5.13** Time-resolved copper plasma emission at different gate delay times of **(A)** 100 ns and **(B)** 2000 ns.

in the early stage of plasma duration, but with the longer gate delay time, peaks with narrow bandwidth appear.

### 5.3.2.4.2 Time-Integrated and Spatially Resolved Spectroscopy

The purpose of this time-integrated and spatially resolved spectroscopy is to measure the LIP emission at different positions in the plasma. Based on the mechanism of laser beam–metal interaction, the emission signals measured along the axial direction from the target surface would show different quantities for different species. A block diagram of the experimental arrangement for time-integrated spatially resolved spectroscopy is shown in Figure 5.14. The details of the system have been described in detail in the literature.[48,53] The metal samples were placed in a T-type stainless steel chamber which could be evacuated with a roughing vacuum pump and filled with various ambient gases such as argon, nitrogen, neon, helium, air, and so on. The LIP emission was imaged onto the entrance slit of the monochromator with a focusing lens. The monochromator and focusing lens were mounted on a stainless steel, multiaxis stage, allowing movement by micrometer positioner that, in turn, allowed the plasma emission to be monitored at different positions on the sample surface.

Lee et al.[48] looked at the LIP caused by the ArF Laser on copper and lead. The irradiance was on the order of $1.5 \times 10^9$ W/cm$^2$. They observed that the appearance and spectra characteristics of the plasma were quite different depending on the target species as well as the position in the plasma. Their further research[53] concentrated on the controlled atmospheric effect in an LIP using

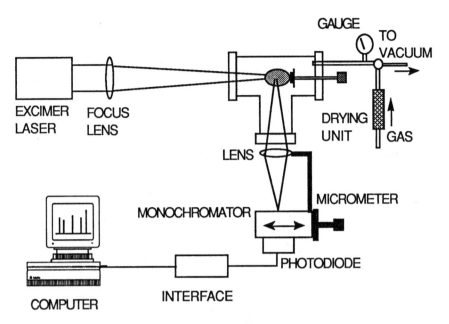

**Figure 5.14**  Schematic diagram of the experimental setup for time-integrated spatially resolved LIBS. From reference 53.

time-integrated and spatially resolved studies. They observed that the plasma consists of two distinct regions when the pressure is reduced below 50 Torr in an air or argon atmosphere. One is near the target surface and is called the *inner sphere plasma.* It emits a strong signal of copper and continuum background. The other is called the *outer sphere plasma.* It surrounds the inner sphere and gives the brilliant blue-green emission of copper with a relatively low background continuum. At pressures lower than 1 Torr, the edge of the outer sphere plasma disappears, but under a helium atmosphere, two distinct plasmas are observed at a pressure of 760 Torr. The highly ionized line is emitted close to the target surface, and the atomic lines are generated in the deeper regions of the plasma. The ion concentration depends on the position in the plasma and reflects the local temperature in the plasma. Typical spectra of LIP copper emissions formed in inner and outer sphere regions are shown in Figure 5.15. As can be seen, the several copper ion lines [Cu(II) and Cu(III)] and background come from the inner sphere plasma, while in the outer sphere plasma, only very clean copper atomic lines [Cu(I)] (510.6, 515.3, 521.8, and 529.3 nm) are observed. The copper plasma emission lines monitored in this experiment are given in Table 5.1. The assignments of the emission lines were based on reference values.[103]

More recently, Mao et al.[104] used time-integrated emission spectroscopy to characterize the emission spectra and excitation temperature spatial profiles within an LIP from a copper target as a function of laser power density. This

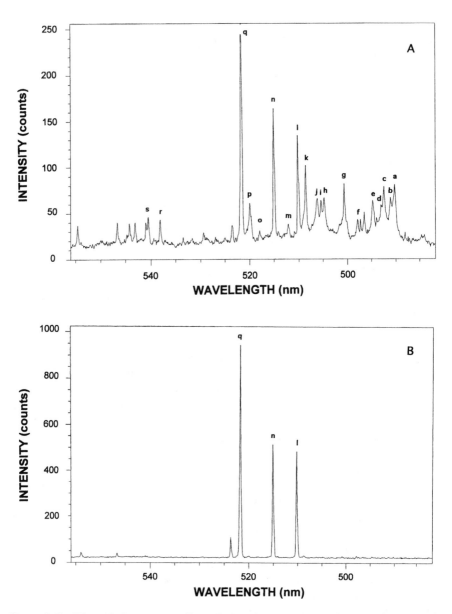

**Figure 5.15**    The emission spectra of laser-induced copper plasma formed in inner sphere region (0.4 mm from the surface) **(A),** and outer sphere region (5.6 mm from the surface) **(B).**

**Table 5.1**  Wavelength and Chemical Species of Copper Emission Lines
Monitored in Laser-Induced Plasma

| Letter | Wavelengh (nm) | Chemical species | Letter | Wavelengh (nm) | Chemical species |
|--------|----------------|------------------|--------|----------------|------------------|
| a | 490.97 | Cu(II) | k | 508.27 | Cu(II) |
| b | 491.84 | Cu(II) | l | 510.55 | Cu(I) |
| c | 492.64 | Cu(II) | m | 512.45 | Cu(II) |
| d | 493.17 | Cu(II) | n | 515.32 | Cu(I) |
| e | 494.30 | Cu(II) | o | 518.34 | Cu(II) |
| f | 498.55 | Cu(II) | p | 520.83 | Cu(III) |
| g | 500.98 | Cu(II) | q | 521.82 | Cu(I) |
| h | 504.74 | Cu(II) | r | 536.98 | Cu(III) |
| i | 505.89 | Cu(II) | s | 541.85 | Cu(III) |
| j | 506.70 | Cu(II) | | | |

research shows how the measured axial spatial emission intensity of the expanding plasma can be influenced by time integration. They concluded that the ability to accurately measure spatial emission intensity and temperature behavior was related to the integration time vs. plasma expansion velocity.

## 5.4  Applications

### 5.4.1  Analysis of Gas

The use of LIBS for the analysis of gas has been pioneered by Radziemski and Cremers and their coworkers.[93,94,105–107] Radziemski and Loree[93] applied LIBS for analyzing trace metals in air. Beryllium in atmospheric pressure air has been detected at 0.7 $\mu g/m^3$, which is 0.6 ng/g of air with a relative standard deviation (RSD) of 30%. The detection limits in air have also been established for Na (0.006 $\mu g/g$), P (1.2 $\mu g/g$), As (0.5 $\mu g/g$), and Hg (0.5 $\mu g/g$). Further work by Ottesen et al.[105–107] developed real-time LIBS to provide in situ determination of elemental composition of particles in a combustion environment. Cremers et al.[94] detected chlorine and fluorine in air directly. Minimum detectable concentrations of chlorine and fluorine in air are 8 and 38 ppm (w/w), respectively. Minimum detectable masses of chlorine and fluorine are, respectively, 80 ng and 2000 ng in air, and 3 ng for both atoms in He. The precision for replicate sample analysis is 8% RSD.

Zhang et al.[108] performed LIBS technique to evaluated its application for practical environments. The relative concentrations of Ca, Fe, Al, Ti, and Si (1:0.36:0.37:0.049:0.017) were inferred by fitting the observed coal-fired flow facility LIBS spectra with computer-simulated spectra.

Recently, Nordstrom[109] studied the LIP emission of $N_2$ and $O_2$, and ambient air in the region of 350 nm to 950 nm. Analysis of air spectra collected by the

LIBS technique indicates that, over a large range of $CO_2$ laser pulse energy, all spectra can be corrected with the use of a simple multiplicative scaling factor. This capability indicates that changes in relative spectral contributions from oxygen and nitrogen are not occurring.

### 5.4.2 Analysis of Liquids

Liquid can be analyzed by LIBS on or under the surface. The advantages of performing LIBS in bulk liquid samples below the surface are that (1) it avoids splashing of the liquid which can contaminate the optical components and introduce noise into the measurements and (2) a stable, free surface is not needed. The detectability of some elements can be increased by forming the breakdown on the liquid surface. Wachter and Cremers[110] formed the plasma on a flowing solution of uranium in 4 M nitric acid. The spectra are shown in Figure 5.16. The advantage of surface excitation is that the solution can be opaque and contain high particle concentrations without significant perturbation of plasma formation.

Initiation of plasma on particles in liquid was investigated by Kitamori et al.[111] They used an Nd:YAG laser (1064 nm, 150 mJ/pulse) focused to $2 \times 10^{13}$ W/cm$^2$ on $CaCO_3$ microparticles (smaller than 0.2 μm) in water. In the laser breakdown emission spectrum, the emission lines of Ca(I) and Ca(II) appeared, however, those features did not appear with aqueous (particle-free) $Ca^{2+}$ solutions. Hence, they concluded that the method was effective for analysis of microparticulate impurities in liquid. Poulain and Alexander[112] applied LIBS to measure the salt concentration in seawater aerosol droplets. A KrF excimer laser ($\lambda$ = 248 nm, 20-ns pulse duration) was used to produce laser-induced breakdown plasmas in aerosol spray. Calibration curves were presented relating the Na(I) 589-nm to $H_\alpha$ 656.3-nm intensity ratio as a function of Na concentration, ranging from 100 to 10,000 ppm. Detection limits of Na by this method were estimated to be approximately 165 ppm for monodisperse sprays and 925 ppm for one case involving a polydisperse spray. Detection of Pb-, Cd-, and Zn-containing aerosols, generated in the laboratory with a nebulizer/heat chamber, was discussed by Essien et al.[113] The detection limits in air were 0.20 ± 0.02 μg/g (Pb), 0.018 ± 0.002 μg/g (Cd), and 0.23 ± 0.14 μg/g (Zn).

Aragon et al.[114] applied the LIBS technique to the determination of carbon content in molten steel samples. An Nd:YAG laser (200 mJ/pulse, 8-ns pulse duration, 1064 nm) was used in this experiment. The precision obtained was 10% for carbon contents in the range of 150–1100 ppm, and the detection limit was 250 ppm.

### 5.4.3 Analysis of Solids

Many samples come to the laboratory as solids and often must be put into solution prior to being analyzed on the instruments. This dissolution process is not only time consuming but also a source of potential analytical error. This

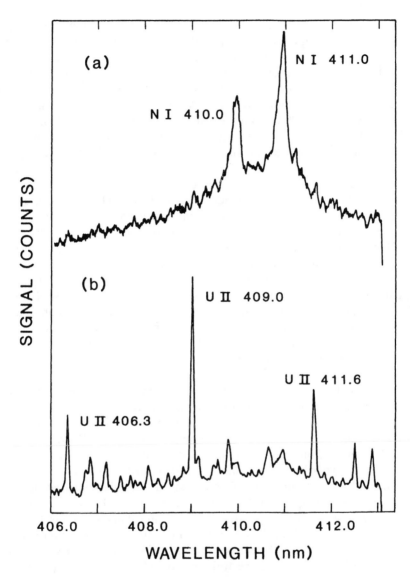

**Figure 5.16**   Spectra of uranium solutions from 406.0 to 413.0 nm for **(a)** 0.1 g/L and **(b)** 10 g/L solution. The scale of **(a)** is expanded by a factor of 5.3 with respect to **(b)**. From reference 110.

can come from contamination or from failure to dissolve all of the analyte into the sample. More importantly, the determination of the trace element distribution in a solid cannot be realized after dissolution of the sample.

Sabsabi et al.[115] applied LIBS to perform elemental analysis of aluminum alloy targets. The plasma was generated by focusing a pulsed Nd:YAG laser ($\lambda$ = 1064 nm, 8-ns pulse duration) on the target in air at atmospheric pressure.

The calibration curve had been obtained by the use of aluminum alloy samples for the Mn(I) 403.3-nm, Si(I) 251.6-nm, Cu(I) 327.4-nm, and Mg(I) 285.2-nm lines. These are approximately straight lines up to 1%. The precision of 50 consecutive measurements of Mg, Mn, Cu, and Si was 4% RSD for the strongest line at high concentration, and 6% at low concentration. The detection limits are element dependent but are on the order of 10 µg/g. They also analyzed a copper alloy by LIBS.[116] Calibration curves for iron, nickel, and silver were produced. The precision ranged from 2 to 10% of the analyte concentration. The detection limits were 20, 10, and 1 µg/g for iron, nickel, and silver, respectively.

An application of LIBS to analyze steels for carbon content was discussed by Aguilera et al.[117] They used a standard Nd:YAG laser (100–300 mJ/pulse, 7-ns pulse duration) on solid samples of stainless and nonstainless steels, generating the plasma in a nitrogen atmosphere. Samples contained up to 1.3% carbon. Good calibration curves were obtained using the ratio of C(I) at 193 nm to Fe(II) at 201.1 nm. Precision of 1.6% and a detection limit of 65 ppm were reported. A subsequent study for the detection of sulfur content in steel samples was carried out by LIBS with the use of a Q-switched Nd:YAG laser (200 mJ/pulse, 7-ns pulse duration) by Gonzalez et al.[118] Calibration curves obtained at a 2-µs delay time from the laser shot were linear in the studies' concentration range. The detection limit of 70 ppm and a precision of 7% were obtained and no noticeable matrix effects were observed.

Lee and Sneddon[119] investigated the quantitative analysis of chromium in standard low-alloy steel samples by LIBS with the use of an ArF excimer laser (100 mJ/pulse, 11-ns pulse duration). They developed a calibration curve that related the chromium concentration in a solid steel matrix to the intensity ratio of Cr(I) at 520.84 nm to Fe(I) at 516.75 nm. The determined chromium concentration ranged from 0.062 to 1.31%. A detection limit of 20 µg/g was obtained. Further work by Lee and Sneddon[120] described a rapid, direct, and sensitive method for determination of potassium in solid glasses based on LIBS technique. This technique allows the direct detection of 0.13 µg/g potassium in solid glass without sample preparation. Results show that the excimer LIBS can provide a precision of about 10% or better, and a linear response up to at least 461 µg. Figure 5.17 shows the emission spectra of LIP with three glass-matrix standard NIST samples over a spectral range of 660–780 nm. The spectra were detected at 0.6 mm from the target surface by time-integrated spatially resolved spectrometry.

Thiem et al.[121] investigated LIBS in ultrahigh vacuum for simultaneous quantitative elemental analysis of NIST transition metal alloy samples. Linear calibration curves were obtained for the elements (in % composition), Al (0.2–1.2%), Cu (0.021–0.49%), Fe (4.5–51.0%), Ni (30.8–80.3%), and Zn (6–12.8%) with the use of nonresonance lines. The detection limits (signal/background = 3) vary with sample composition complexity and range from 0.0001% for Ni in a sample copper alloy (SRM1111) to 0.16% for Al in a complex granular sample (SRM 349a). Kurniawan et al.[122] carried out the analysis of glass samples by

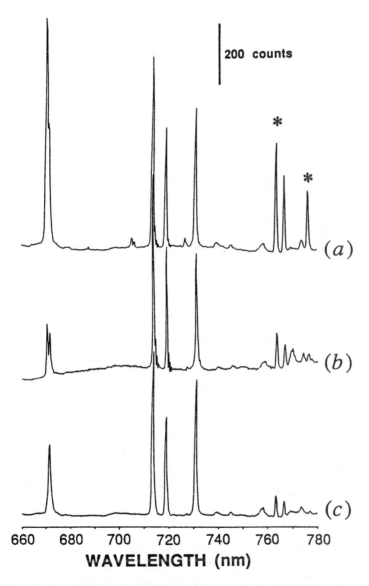

**Figure 5.17**   Excimer LIBS spectra of glass-matrix standards (NIST 610 series) obtained using a single laser shot with energy of 150 mJ/pulse. The 764.6-nm line for K(I) was used for calibration. Standards: **(a)** NIST 610, 461 µg/g; **(b)** NIST 612, 64 µg/g; **(c)** NIST 614, 30 µg/g. From reference 120.

XeCl excimer laser and TEA $CO_2$ laser. Initial quantitative analysis was performed on a number of glass samples, and a linear calibration curve with a slope of near unity was obtained at 1 Torr of ambient gas. Furthermore, light elements such as Li and B, which are usually difficult to observe with an X-ray fluorescence method, were also successfully detected with a very low detection limit of less than 10 ppm. Various other elements, such as Na, Mg, Al, K, Ca, Ti, Zn, Zr, and Ba were also detected. The detection limits of these elements were much lower than those usually required for glass analysis. Recently, Jensen et al.[123] examined a number of factors that influence the use of LIBS to detect the cation content of model environmental samples ($Eu_2O_3$ and $K_2Cr_2O_7$ mixture). They suggest that LIBS at ambient atmospheric pressure has some promise for the direct generation of quantitative data on these environmental samples and considerable promise as a fast screening tool for heavy metals.

Anderson et al.[124] conducted an application of LIBS for depth profile measurements of coatings on steel. Linear calibrations against coating thickness for Zn/Ni (2.7 to 7.2 µm) and Sn (0.38 to 1.48 µm) on steel were achieved with good precision (3.5% RSD). An ultrathin coating of Cr (20 nm) on steel was also detected by this technique. Hakkanen et al.[125] applied LIBS to measure the coating coverage, coating weight distribution, and three-dimensional distribution of elements in paper coatings. The experimental results were used to evaluate the quality of the paper coating. The coating pigment was kaolin, and the atomic emission of silicon at 251 nm was used to monitor the pigment content. They used an XeCl excimer laser (0.2 mJ/pulse, 10-ns pulse duration) and detected the silicon and calcium atomic emission lines. Remote elemental analysis by LIBS using a fiber-optic cable was reported recently by Cremers et al.[126] The laser was operated at 10 Hz and the pulse energy incident on the input end of the fiber was 100 mJ with up to 84 mJ incident on the sample because of recent advances in fiber-optic materials.[127] The content of Ba and Cr in soils was determined by remote LIBS. The detection limits for Ba and Cr were 26 and 50 ppm, respectively, and the RSDs ranged from 6% to 20%. This remote LIBS technique opens up many new areas of application including analysis in hostile environments.

## 5.5 Conclusions

Laser-induced breakdown spectrometry (LIBS) is a laser-based, (mostly) non-destructive, sensitive optical analytical technique used to detect certain atomic, ionic, and molecular species in solid, liquid, and gas. Some of the newest applications of LIBS are likely to be active topics of investigation for years to come. The primary advantages of LIBS are that it can be used for the direct, in situ, determination of elements in samples without any pretreatment, and it can be applied, when speed is more important than accuracy, to on-line analysis. Therefore it can be used to classify geological materials and chemical products, control product quality, and so forth. Although many of the methods based on laser ablation techniques currently being developed have not achieved satisfactory (compared to more traditional methods) detectability, accuracy, and precision, they have stimulated experimentation that will hopefully someday bring about newer and better analytical instrumentation based on lasers for the analytical chemist.

## Acknowledgments

The authors gratefully acknowledge the partial financial support from the Korea Science and Engineering Foundation (KOSEF) (Grant No. 951-0304-050-2).

## References

1. *Laser-Induced Plasmas and Applications;* Radziemski, L.J.; Cremers, D.A. Eds.; Marcel Dekker: New York, 1989.

2. Darke, S.A.; Tyson, J.F. *J. Anal. Atom. Spectrom.* **1993,** *8,* 145.

3. Radziemski, L.J. *Microchem. J.* **1994,** *50,* 218.

4. Sneddon, J. *Spectroscopy* **1994,** *9,* 34.

5. Thiem, T.L.; Lee, Y.I.; Sneddon, J. *Microchem. J.* **1992,** *45,* 1.

6. Majidi, V. *Spectroscopy* **1993,** *8,* 16.

7. Russo, R.E. *Appl. Spectrosc.* **1995,** *49,* 14A.

8. Thiem, T.L.; Lee, Y.I.; Sneddon, J. In *Advances in Atomic Spectroscopy;* Sneddon, J., Ed.; JAI Press: London, 1995; Vol. 2, p. 179.

9. Moenke-Blankenburg, L. *Laser Microanalysis;* Wiley: New York, 1989.

10. Grey-Morgan, C. *Rep. Prog. Phys.* **1975,** *38,* 621.

11. Ready, J.F. *Effect of High Power Laser Radiation;* Academic: New York, 1971.

12. Anisimov, S.I. *Sov. Phys. JETP* **1968,** *27,* 182.

13. Anisimov, S.I.; Melshantsev, B.I. *Sov. Phys. Solid State* **1973,** *15,* 743.

14. Caruso, A.; Bertotti, B.; Giupponi, P. *Nuovo Cimento* **1966,** *XIV B,* 176.

15. Weyl, G.M. In *Laser-Induced Plasmas and Applications;* Radziemsk, L.J.; Cremers, D.A., Eds.; Marcel Dekker: New York, 1989; Chapter 1.

16. Weyl, G.; Pirri, A.; Root, R. *AIAA J* **1981,** *19,* 460.

17. Maher, W.E.; Hall, R.B.; Johnson, R.R. *Appl. Phys.* **1974,** *45,* 2138.

18. Steverding, B. *J. Appl. Phys.* **1974,** *45,* 3507.

19. Pirri, A.; Root, R.G.; Wu, P.K.S. *AIAA J.* **1978,** *16,* 1296.

20. Shah, P.; Armstrong, R.L.; Radziemski, L.J. *J. Appl. Phys.* **1991,** *1,* 180.

21. Cottet, F.; Romain, J.P. *Phys. Rev. A* **1981,** *25,* 576.

22. Balazs, L.; Gijbels, R.; Vertes, A. *Anal. Chem.* **1991,** *63,* 314.

23. Eloy, J.F. *Recent Developments in Laser Microprobe Mass Spectrometry, Proceedings of the 5th International Symposium, Dresden;* 1980; Vol. 2, p96.

24. Bingham, R.A.; Salter, P.L. *Anal. Chem.* **1976,** *48,* 1735.

25. Fabbro, F.; Fabre, E.; Amiranoff, F.; Garbeau-Labaune, C.; Virmont, J.; Weinfield, M.; Max, C.E. *Phys. Rev. Ser. A* **1980,** *26,* 2289.

26. Yaakobe, B.; Boehly, T.; Bourke, P.; Conturie, Y.; Craxton, R.S.; Delettrez, J.; Forsyth, J.M.; Frankel, R.D.; Goldman, L.M.; McCrory, R.L.; Richardson, M.C.; Seka, W.; Shvarts, D.; Soures, J.M. *Opt. Commun.* **1981,** *39,* 175.

27. Ripin, B.H.; Decoste, R.; Obenschain, S.P.; Bodner, S.E.; EcLean, E.A.; Young, F.C.; Whitlock, R.R.; Armstrong, C.M.; Grun, J.; Stamper, J.A.; Gold, S.H.; Nagal, D.J.; Lehmberg, R.H.; McMahon, J.M. *Phys. Fluids* **1980,** *23,* 1012.

28. Maaswinkel, A.G.M.; Eidmann, K.; Sigel, R.; Witkowdki, S. *Opt. Commun.* **1984,** *51,* 255.

29. Kwok, H.S.; Mattocks, P.; Shi, L.; Wang, X.W.; Witanachchi, S.; Ying, Q.Y.; Zheng, J.P.; Shaw, D.T. *Appl. Phys. Lett.* **1988,** *52,* 1995.

30. Kagawa, K.; Kawai, K.; Tani, M.; Kobayashi, T. *Appl. Spectrosc.* **1994,** *48,* 198.

31. Kurniawan, H.; Nakajima, S.; Batubara, J.E.; Marpaung, M.; Okamoto, M.; Kagawa, K. *Appl. Spectrosc.* **1995,** *49,* 1067.

32. Lincolin, K.A. *Int. J. Mass. Spectrom. Ion. Phys.* **1974,** *13,* 45.

33. Selter, K.P.; Kunze, H.J. *Phys. Scrip.,* **1982,** *54,* 879.

34. Opauszky, I. *Pure Appl. Chem.* **1982,** *54,* 879.

35. Carroll, P.K.; Kennedy, E.T. *Contemp. Phys.* **1981,** *22,* 61.

36. Dyer, P.E. *Appl. Phys. Lett.* **1989,** *55,* 1630.

37. Vertes, A.; De Wolf, M.; Juhasz, P.; Giijbels, R. *Anal. Chem.* **1989,** *61,* 1029.

38. Hwang, Z.W.; Teng, Y.Y.; Li, K.P.; Sneddon, J. *Appl. Spectrosc.* **1991,** *45(3),* 435.

39. Sneddon, J.; Li, K.P.; Hwang, Z.W.; Teng, Y.Y. *Proceedings of the International Conference on LASERS 1990;* STS Press: McLean, VA, 1991; p. 648.

40. Hwang, Z.W.; Teng, Y.Y.; Li, K.P.; Sneddon, J. *Spectrosc. Lett.* **1991,** *24(9),* 1173.

41. Sneddon, J. *Proceedings of the International Conference on LASERS 1989;* STS Press: McLean, Virginia, 1990; p. 750.

42. Iida, Y.; Yeung, E.S.; *Appl. Spectrosc.* **1994,** *48,* 945.

43. Allemand, C.D. *Spectrochim. Acta B* **1972,** *27,* 185.

44. Ishizuka, T. *Anal. Chem.* **1973,** *45,* 538.

45. Prochorov, A.M.; Batanov, V.A.; Bunkin, F.V.; Fedorov, V.B. *IEEE J. Quantum Electron.* **1973,** *QE-9,* 503.

46. Dimitrov, G.; Maximova, Ts. *Spectrosc. Lett.* **1981,** *14,* 734.

47. Dimitrov, G.; Zheleva, Ts. *Spectrochim. Acta B* **1984,** *39,* 1209.

48. Lee, Y.I.; Sawan, S.P.; Thiem, T.L.; Teng, Y.Y.; Sneddon, J. *Appl. Spectrosc.* **1992,** *46,* 436.

49. Lee, Y.I.; Sneddon, J. *Microchem. J.* **1994,** *50,* 235.

50. Iida, Y. *Appl. Spectrosc.* **1989,** *43,* 229.

51. Iida, Y.; Morikawa, H.; Tsuge, A.; Uwamino, Y.; Ishizuka, T. *Anal. Sci.* **1991,** *2,* 61.

52. Grant, K.J.; Paul, G.L. *Appl. Spectrosc.* **1990,** *44,* 1349.

53. Lee, Y.I.; Thiem, T.L.; Kim, G.H.; Teng, Y.Y.; Sneddon, J. *Appl. Spectrosc.* **1992,** *46(11),* 1597.

54. (a) Lee, Y.I.; Song, K.; Cha, H.K.; Lee, C.M.; Park, M.C.; Lee, G.H. Presented at the PittCon '96, Chicago, IL, 1996; paper 216P. (b) Song, K.; Cha, H.K.; Lee, J.M.; Lee, Y.I. *Appl. Spectros.* submitted for publication.

55. Kagawa, K.; Kawai, K.; Tani, M.; Kobayashi, T. *Appl. Spectrosc.* **1994,** *48,* 198.

56. Mao, X.L.; Chan, W.T.; Shannon, M.A.; Russo, R.E. *J. Appl. Phys.* **1993,** *74,* 4915.

57. Kuzuya, M.; Matsumoto, H.; Takechi, H.; Mikami, O. *Appl. Spectrosc.* **1993,** *47,* 1659.

58. Goode, S.R.; Pipes, D.T. *Spectrochim. Acta* **1981,** *36B,* 925.

59. Klueppel, R.J.; Walters, J.P. *Spectrochim. Acta* **1980,** *35B,* 431.

60. Majidi, V.; Coleman, D.M. *Appl. Spectrosc.* **1987,** *41,* 200.

61. Albers, D.; Johnson, E.; Tisak, M.; Sacks, R. *Appl. Spectrosc.* **1986,** *40,* 60.

62. Albers, D.; Sacks, R. *Spectrochim. Acta* **1986,** *41B,* 391.

63. Albers, D.; Tisak, M.; Sacks, R. *Appl. Spectrosc.* **1987,** *41,* 131.

64. Johnson, E.T.; Sacks, R.D. *Anal. Chem.* **1987,** *59,* 2170.

65. Johnson, E.T.; Sacks, R.D. *Anal. Chem.* **1987,** *59,* 2176.

66. Johnson, E.T.; Sacks, R.D. *Appl. Spectrosc.* **1988,** *42,* 77.

67. Hontzopoulos, E.; Charalambidis, D.; Fotakis, C.; Farkas, G.; Horvath, Z.; Toth, C. *Opt. Commun.* **1988,** *67,* 124.

68. Kumar, V.; Thareja, R.K.; *J. Appl. Phys.* **1988,** *64,* 5269.

69. Mason, K.J.; Goldberg, J.M.; *Appl. Spectrosc.* **1991,** *45,* 370.

70. Mason, K.J.; Goldberg, J.M. *Anal. Chem.* **1987,** *59,* 1250.

71. Mason, K.J.; Goldberg, J.M. *Appl. Spectrosc.* **1991,** *45,* 1444.

72. Dirnberger, L.; Dyer, P.E.; Farrar, S.R.; Key, P.H. In *American Institute of Physics Conference Proceedings,* New York, 1994; p 349.

73. Kirbright, G.F.; Sargent, M.; Vetter, S. *Spectrochim. Acta* **1970,** *25B,* 465.

74. Hood, W.H.; Niemczyk, T.M. *Appl. Spectrosc.* **1987,** *41,* 674.

75. Mehs, D.M.; Niemczyk, T.M. *Appl. Spectrosc.* **1981,** *35,* 66.

76. Faires, L.M.; Palmer, B.A.; Englemen, R., Jr.; Niemczyk, T.M. *Spectrochim. Acta* **1984,** *39B,* 819.

77. Uchida, H.; Tanbe, K.; Nojiri, Y.; Haraguchi, H.; Fuwa, K. *Spectrochim. Acta* **1981,** *36B,* 711.

78. Kalnicky, D.J.; Fassel, V.A.; Kniseley, R.N. *Appl. Spectrosc.* **1977,** *31,* 137.

79. Kagawa, K.; Ohtani, M.; Yokoi, S.; Nakajima, S. *Spectrochim. Acta* **1984,** *39B,* 525.

80. Ursa, I.; Stoica, M.; Mihailescu, N.; Hening, A.I.; Prokhorov, A.M.; Nikitin, P.I.; Knoov, V.I.; Silenok, S.J. *J. Appl. Phys.* **1989,** *66,* 5204.

81. Radziemski, L.J.; Cremers, D.A.; Niemczyk, T.M. *Spectrochim. Acta* **1985,** *40B,* 517.

82. Cremers, D.A.; Radziemski, L.J.; Loree, T.R.; Hoffman, N.M. *Anal. Chem.* **1983,** *55,* 1246.

83. Fabbro, F.; Fabre, E.; Amiranoff, F.; GArbeau-Labaune, C.; Virmont, J.; Weinfield, M.; Max, C.E. *Phys. Rev. Ser. A* **1980**, *26*, 2289.

84. Evtushenki, T.P.; Ostrovskaya, G.V. *Sov. Phys. Tech. Phys.* **1970**, *15*, 823.

85. Adamson, A.W.; Cimolino, M.C. *J. Phys. Chem.* **1984**, *88*, 488.

86. Laporte, P.; Damany, N.; Damany, H. *Opt. Lett.* **1987**, *12*, 987.

87. Carroll, P.K.; Kennedy, E.T.; O'Sullivan, G. *Opt. Lett.* **1978**, *2*, 72.

88. Majidi, V.; Ratliff, J.; Owens, M. *Appl. Spectrosc.* **1991**, *45*, 473.

89. Xu, N.; Majidi, V. *Appl. Spectrosc.* **1993**, *47*, 1134.

90. Simeonsson, J.B.; Miziolek, A.W. *Appl. Opt.* **1993**, *32*, 939.

91. Lee, Y.I. Ph.D. Dissertation, University of Massachusetts, Lowell, MA, 1993.

92. Carroll, P.K.; Kennedy, E.T.; *Contemp. Phys.* **1981**, *22*, 61.

93. Radziemski, L.J.; Loree, T.R.; Cremers, D.A.; Hoffman, N.M. *Anal. Chem.* **1983**, *55*, 1246.

94. Cremers, D.A.; Radziemski, L.J. *Anal. Chem.* **1983**, *55*, 1252.

95. Autin, M.; Briand, A.; Mauchien, P. *Spectrochim. Acta* **1993**, *48B*, 851.

96. Leis, F.; Sdorra, W.; Ko, J.B.; Niemax, K. *Mikrochim. Acta* **1989**, *II*, 185.

97. Kagawa, K.; Yokoi, S. *Spectrochim. Acta* **1982**, *37B*, 789.

98. Kagawa, K.; Matsuda, Y.; Yokoi, S.; Nakajima, S. *J. Anal. Atom. Spectrom.* **1988**, *3*, 415.

99. Kagawa, K.; Ohtani, M.; Yokoi, S.; Nakajima, S. *Spectrochim. Acta* **1984**, *39B*, 525.

100. Kagawa, K.; Tani, M.; Ueda, H.; Sasaki, M.; Mizukami, K. *Appl. Spectrosc.* **1993**, *47*, 1562.

101. Piepmeier, E.H.; Osten, D.E. *Appl. Spectrosc.* **1971**, *25*, 642.

102. Joseph, M.R.; Zu, N.; Majidi, V. *Spectrochim. Acta* **1994**, *49B*, 89.

103. Reader, J.; Corliss, C.H.; Wiese, W.L.; Martin, G.A. *Wavelengths and Transition Probabilities for Atoms and Atomic Ions;* U.S. Department of Commerce. National Institute of Standards and Technology. U.S. Government Printing Office: Washington DC, 1980.

104. Mao, X.L.; Shannon, M.A.; Fernandez, A.J.; Russo, R.E. *Appl. Spectrosc.* **1995**, *49*, 1054.

105. Ottesen, D.K.; Wang, J.C.F.; Radziemski, L.J. *Appl. Spectrosc.* **1989**, *43*, 1967.

106. Ottesen, D.K. "Laser-Spark Spectroscopy for In Situ Analysis of Particles in Combustion Flows"; Sandia National Laboratories Technical Report; Sandia National Laboratory: Albuquerque, NM, 1988; pp. 8–14.

107. Ottesen, D.K.; Baxter, L.L.; Baxter, L.J.; Radziemsk, L.J.; Burrows, L.F. *Energy Fuels* **1991**, *5*, 304.

108. Zhang, H.; Singh, J.P.; Yueh, F.-Y.; Cook, R.L. *Appl. Spectrosc.* **1995**, *49*, 1617.

109. Nordstrom, R.J. *Appl. Spectrosc.* **1995**, *49*, 1490.

110. Wachter, J.R.; Cremers, D.A. *Appl. Spectrosc.* **1987**, *41*, 1042.

111. Kitamori, T.; Matsui, T.; Sakagami, M.; Sawada, T. *Chem. Lett.* **1989**, 2205.

112. Poulain, D.E.; Alexander, D.R. *Appl. Spectrosc.* **1995**, *49*, 569.

113. Essien, M.; Radziemski, L.J.; Sneddon, J. *J. Anal. Atom. Spectrosc.* **1988**, *3*, 985.

114. Aragon, C.; Aguilera, J.A.; Campos, J. *Appl. Spectrosc.* **1993**. *47*, 606.

115. Sabsabi, M.; Cielo, P. *Appl. Spectrosc.* **1995,** *49,* 499.

116. Sabsabi, M.; Cielo, P. *J. Anal. Atom. Spectrom.* **1995,** *10,* 643.

117. Aguilera, J.A.; Aragon, C.; Campos, J. *Appl. Spectrosc.* **1992,** *46,* 581.

118. Gonzalez, A.; Ortiz, M.; Campos, J. *Appl. Spectrosc.* **1995,** *49,* 1633.

119. Lee, Y.I.; Sneddon, J. *Spectrosc. Lett.* **1992,** *25,* 881.

120. Lee, Y.I.; Sneddon, J. *Analyst* **1994,** *119,* 1441.

121. Thiem, T.L.; Salter, R.H.; Gardner, J.A.; Lee, Y.I.; Sneddon, J. *Appl. Spectrosc.* **1994,** *48,* 58.

122. Kurniawan, H.; Nakajima, S.; Batubara, J.E.; Marpaung, M.; Okamoto, M.; Kagawa, K. *Appl. Spectrosc.* **1995,** *49,* 1067.

123. Jensen, L.C.; Langfor, S.C.; Dickinson, J.T.; Addleman, R.S. *Spectrochim. Acta* **1995,** *50B,* 1501.

124. Anderson, D.R.; McLeod, C.W.; English, T.; Smith, A.T. *Appl. Spectrosc.* **1995,** *49,* 691.

125. Hakkanen, H.J.; Korppi-Tommola, J.E.I. *Appl. Spectrosc.* **1995,** *49,* 1721.

126. Cremers, D.A.; Barefield, J.E., II; Koskelo, A.C. *Appl. Spectrosc.* **1995,** *49,* 857.

127. Trott, W.M.; Meeks, K.D. *J. Appl. Phys.* **1990,** *67,* 3297.

# 6

# Laser-Enhanced Ionization Spectrometry

*David J. Butcher*

## 6.1 Introduction

Laser-enhanced ionization (LEI) spectrometry is a highly sensitive technique for elemental analysis, with detection limits in the part per trillion range. Light from a laser is used to excite atoms (or less commonly, molecules) in a conventional analytical atom cell to a sufficiently high energy level to induce an enhancement in the rate of collisional ionization compared to background rates of ionization. The incorporation of an electric field in the atom cell allows measurement of the ionized species as a current. LEI may be distinguished from other ionization techniques, such as direct laser ionization (DLI), multiphoton ionization (MPI) and resonance ionization spectroscopy (RIS) because these other techniques involve laser photoionization. DLI and MPI are performed in analytical atom cells, while RIS is usually performed in low-pressure vapor cells.

This chapter includes a summary of previous reviews of the technique, followed by a description of its fundamental principles. Instrumentation for LEI is described, with particular emphasis on the role of the laser. Interferences that limit the use of LEI for real sample analysis are described, along with methods to reduce their magnitude. Major analytical developments in LEI are summarized since a previous major review by Green and Seltzer[1] (1989 to the present). A brief conclusion section summarizes the status of LEI as a method for elemental analysis. Photoionization techniques have been omitted from this chapter, as well as diagnostic studies of plasmas by LEI.

## 6.2 Previous Reviews

This section summarizes major reviews of LEI[1-8] since 1986. Green and Seltzer[1] described the principles of LEI, DLI, and MPI, including signal production, detection, and instrumentation. Axner and Rubinsztein-Dunlop[2] reviewed flame LEI and discussed theoretical expressions for the signal, its analytical performance, and interferences. Axner et al.[3] reviewed the analytical performance of LEI in flames and graphite furnaces. Kuzyakov and Zorov[4] comprehensively reviewed LEI and photoionization spectrometry techniques, including DLI, MPI, RIS, and laser resonance ionization mass spectrometry (RIMS). Turk[5] reviewed LEI in flames and plasmas, including a comprehensive table of the best detection limit for each element. Sjöström and Mauchien,[6] Niemax,[7] and Omenetto[8] reviewed the analytical performance and potential of a variety of laser atomic spectrometric techniques, including LEI.

## 6.3 Fundamental Principles

Four physical processes must occur to obtain an LEI signal; (1) atomization of the sample, which (in most cases) involves conversion of a liquid sample to gaseous analyte atoms; (2) excitation of the ground-state atoms to an excited state with laser radiation; (3) ionization of the excited state atoms; and (4) collection of the resulting charged species (ions and electrons).[1-5,9] Optimum sensitivity for LEI is obtained when the efficiency of each of these steps approaches 100%. The degree of atomization in LEI, as in other atomic techniques, is dependent upon the atom cell, and hence is the same as in commonly used atomic absorption or emission techniques.[10]

The excitation and ionization steps, which are illustrated in Figure 6.1, are discussed together because the choice of excitation wavelength is dependent upon the available atomic transitions and the ionization limit. A laser source is required for LEI because conventional sources (e.g., hollow cathode lamps and electrodeless discharge lamps) have insufficient intensity to significantly populate an excited energy level and enhance the rate of ionization. It is desirable to excite the atoms to an excited state near the ionization limit because the rate of collisional ionization increases by 100 times per electron volt above the ground state. However, in order to obtain maximum sensitivity, it is also necessary to populate the excited state with the maximum possible fraction of atoms, a condition that is called optical saturation. A graph of laser pulse energy versus LEI signal shows a linear dependence at low pulse energies, but at higher pulse energies, a maximum LEI signal is obtained (Figure 6.2). Very high laser pulse energies are required to saturate transitions to levels very close to the limit, and therefore, optimum sensitivity is obtained by the selection of a transition that is near the ionization limit but can be saturated with the available laser system. In addition, significant spectral interferences have been observed with high laser

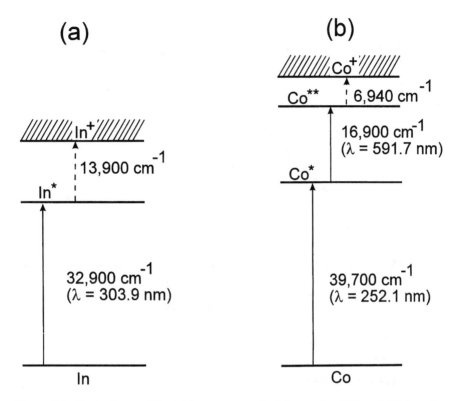

**Figure 6.1** Excitation and ionization processes in (**a**) one-step LEI and (**b**) two-step LEI. Adapted with permission from reference 9.

pulse energies (Section 6.5). The use of one wavelength of laser radiation is referred to as one-step LEI.

Complete ionization efficiency with one wavelength of laser radiation can be obtained for elements (e.g., indium) whose ionization limit is within 7.5 eV of the ground state. However, for elements whose ionization limit exceeds this level (e.g., cobalt), the use of two wavelengths of laser light (two-step LEI) provides improved sensitivity (Figure 6.1). The atoms are excited from the ground state (Co) to an intermediate level (Co*) with one laser photon, followed by excitation to an energy level (Co**) near the ionization limit by a second photon of a different wavelength. A significant increase in the rate of ionization and decrease in detection limit (one to three orders of magnitude) is achieved by the use of two-step LEI. In order to achieve maximum sensitivity, both laser transitions must be saturated. Although two-step LEI provides improved sensitivity for many elements, it is more expensive and difficult to align (Section 6.4.2).

The efficiency of charge collection is dependent upon instrumental design which is discussed in Section 6.4.

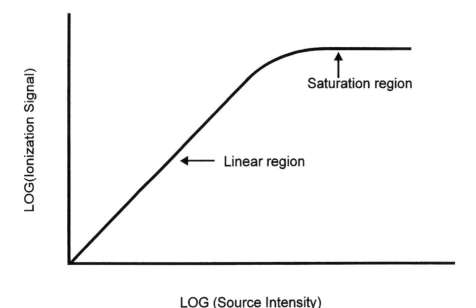

**Figure 6.2**   Saturation curve for LEI.

## 6.4 Instrumentation

A block diagram of instrumentation for LEI is shown in Figure 6.3.[1-5] The system is composed of three components: a laser system, which includes a pump laser, one or two dye lasers, and a frequency doubling system; an atom cell, where the gaseous analyte atoms are produced; and a detection system, consisting of a high-voltage power supply, collection electrodes, a preamplifier, a boxcar integrator, a system to trigger the boxcar, and a computer for instrument control. Each component is discussed in detail below.

### 6.4.1 Laser Systems

The choice of laser system for LEI is governed by four parameters: (1) wavelength coverage, (2) the radiant intensity of the laser light, (3) (for a pulsed laser) the repetition rate, and (4) the laser linewidth.[1-5] For LEI the laser must be able to produce wavelengths throughout the visible and ultraviolet regions of the spectrum. Lasers capable of this wide wavelength coverage include three component laser systems consisting of a pump laser (usually excimer, Nd:YAG, or nitrogen), a dye laser, and a frequency-doubling system. In order to do two-step LEI, a second dye laser is irradiated by the same pump laser to produce a second wavelength for excitation. Pulse energies of 50–100 μJ are required to

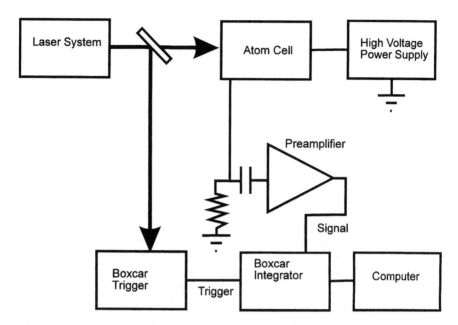

**Figure 6.3**   Schematic diagram of instrumentation for LEI.

saturate the ultraviolet (frequency-doubled) transitions, while 300–1000 µJ may be required in the visible region for the second transition. Excimer and Nd:YAG lasers have been used for the majority of recent LEI work because of their high radiant intensities. Repetition rates of 10–30 Hz are adequate for flame and plasma LEI, in which a continuous signal is obtained due to constant aspiration of sample solution into the atom cell. However, for atom cells that provide a transient signal, such as electrothermal atomizers (graphite furnaces), a repetition rate of a least 60 Hz is required to adequately sample the signal.[11] Typical laser linewidths for LEI are 0.002–0.008 nm.[11] It should be pointed out that continuous wave laser systems have been used for LEI, but their wavelength coverage is limited to the visible region, which greatly restricts the number of determinable elements. Details regarding the mode of operation of the laser systems is given in Chapter 2 of this volume.

   In spite of its high sensitivity, the unreliability and difficulty of use of the laser systems have relegated the use of LEI to laboratory-constructed instruments in a handful of government and academic laboratories. Excimer lasers require relatively frequent and time-consuming repairs. In order to obtain optimum sensitivity for every element to be determined, it is usually necessary to change the dye(s) in the dye laser(s). The development of reliable, easy-to-use laser systems must be achieved before LEI will be routinely used for elemental analysis.

### 6.4.2 Atom Cells

A variety of atom cells have been employed for LEI, including flames, graphite furnaces, and inductively coupled plasmas. The majority of LEI research has been done with flame atomization which is the primary focus of this section.

The most commonly used flame for LEI is air-acetylene equipped with a pneumatic nebulizer for sample introduction.[1-5] Its principal advantages include applicability to a wide variety of elements, high stability, and relatively low background signals. Other flames that have been employed for LEI include nitrous oxide–acetylene and hydrogen-oxygen-argon. Nitrous oxide–acetylene flames are widely employed in atomic absorption spectrometry for the determination of involatile elements (e.g., Al and Cr),[10] but their use has been limited for LEI. Axner and Rubinsztein-Dunlop[12] recently reported the determination of chromium by flame LEI and chose to use the air-acetylene flame. A contamination-limited (reportedly from the burner head or nebulizer) detection limit of 2 ng/mL was obtained, which is remarkably low considering the small fraction of chromium atomized in this flame.

Compared to other flame spectrometry techniques, the most unique aspect of the flame system employed for LEI is the electrode system to collect the laser-induced charges. The various electrode arrangements that have been used for LEI are summarized in Figure 6.4. Generally, electric fields in flame LEI have varied between a few hundred and a few thousand volts. LEI electrodes have been positioned in close proximity to, but outside of the flame (external), or immersed in the flame. The use of external plate electrodes has been shown to provide lower detection limits, although external electrodes have been shown to be more susceptible to interferences induced by easily ionized elements (EIEs) (Section 6.5).

### 6.4.3 Detection System

Components for LEI detection systems have been selected by the characteristics of the LEI signal.[1-5,13,14] With pulsed laser systems, the LEI signal is observed as a current a few hundred nanoseconds after the laser pulse, with a duration of several hundred nanoseconds. The magnitude of the current is on the order of nano- to micro-A. After the optogalvanic signal has been converted to a voltage and preamplified, a boxcar integrator is used to monitor the signal only during the temporal window in which the LEI signal is observed. The boxcar integrator significantly decreases noise levels because analytical measurements are made only during the appropriate window (gate). A small fraction of laser light is allowed to reach a photodetector (photodiode or photomultiplier tube) to trigger the boxcar. The output of the boxcar is sent to a computer for data collection and manipulation.

## 6.5 Interferences

LEI interferences are phenomena that affect the detection limit, linearity, accuracy, and precision of the technique.[1-5,15] They can be classified into two broad

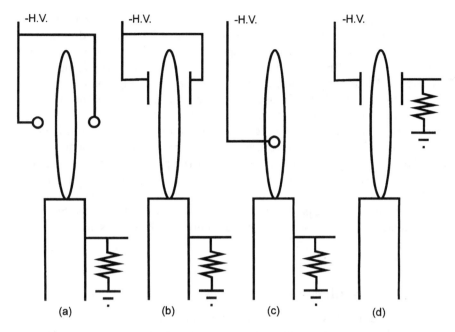

**Figure 6.4** Various electrode arrangements used for LEI spectrometry in flames: (**a**) external rods, (**b**) external plates with collection from the burner, (**c**) immersed water-cooled cathode, (**d**) external plates with collection from the low voltage electrode. The burner head is electrically isolated from ground in all configurations. Adapted with permission from reference 2.

categories: non-laser-induced interferences and laser-induced interferences. The origin of the most severe LEI interferences is the presence of easily ionized elements (EIEs), such as alkali and alkaline earth metals, and hence the conclusion of this section is a discussion of these interferences and methods to alleviate their effects. As in the previous section, discussion will focus on flame LEI in the commonly used air-acetylene flame and on interferences that are unique to LEI. The reader is referred to reference 10 for a general discussion of interferences in flames (e.g., chemical and physical interferences).

### 6.5.1 Non-Laser-Induced Interferences

A number of interferences are present in the absence of laser excitation.[1–5,15] A background signal of a few microamps is observed in flames in the absence of solution aspiration due to ionization of molecules. The use of electrodes that are immersed in the flame induces emission of electrons from the electrode surface. Electrode noise is responsible for the higher detection limits obtained with an immersed signal collection system compared to external plate electrodes. The most troublesome non-laser-induced interference is thermal ionization of matrix species, such as easily ionized elements (EIEs).

EIEs degrade LEI signals by two physical mechanisms: (1) increased flame background current and (2) impaired collection of LEI signals due to space charge effects. Turk and Kingston[15] reported that, while background currents in the absence of EIEs are typically 4–9 µA in an air-acetylene flame using a water-cooled immersed cathode, the addition of 100 µg/mL sodium increased the background to 640 µA. In order to collect charges produced by laser excitation, an electric field must be present in the probe volume of the flame. Electrons have greater mobility than ions and are more quickly removed from the flame than cations. A positive charge, called a space charge, consequently accumulates in the flame around the cathode. EIEs produce a space charge that may occupy a significant volume of the flame, and effectively eliminate the electric field. Laser-induced charges produced in the space charge region are consequently not detected. Space charge effects are considerably reduced by the use of an immersed cathode, which allows detection of LEI signals at sodium concentrations up to 1000 µg/mL.

### 6.5.2  Laser-Induced Interferences

Laser-induced interferences can be separated into the following categories:[1–5,15] (1) spectral interferences, which involve the overlap of the analyte spectral line with that of a matrix element; (2) photoionization, in which a matrix species is directly photoionized from the ground state; (3) laser-enhanced ionization of molecules; (4) laser-enhanced ionization of matrix elements by amplified spontaneous emission; and (5) radio-frequency interference generated by the pump laser that is collected by the detection system. Spectral interferences commonly involve the overlap of alkali metal spectral lines with those of the analyte. Several alkali metal transitions have broad wings that allow excitation of the atoms even at wavelengths several nanometers from the center of the absorption line. The use of two-step LEI has been used to quantify the magnitude of these background signals, because the overlap of a matrix element with both transitions is extremely unlikely. Hence, signal plus background can be measured with both lasers on, and background signals can be estimated by measuring the background with one laser on and the other off, and vice versa. Subtraction of the background measurements from the signal plus background measurement provides a background-corrected signal.

Photoionization interferences are most severe for samples that contain relatively high concentrations of lithium or sodium with the use of analytical wavelengths below 245 nm. Hence, for many elements this interference can be minimized by the use of alternative transitions. However, elements such as cadmium do not have suitable alternative transitions, and photoionization may be a significant interference.

Molecular species may absorb laser light and subsequently ionize, producing a background signal. Examples of molecules that produce background signals include NO, YO, LaO, and alkaline earth oxides (BaO, CaO). The use of an alternative wavelength, if possible, is probably the best approach to minimize these interferences.

Amplified spontaneous emission (ASE) is relatively low-energy broadband emission from a dye laser that may be absorbed by matrix elements and produce a background signal. ASE has been significantly reduced in modern dye laser systems. Radiofrequency interferences involve the transmission of radiofrequency noise produced by the pump laser to the detection system. These interferences are generally minimized by enclosing the flame and detection system in a Faraday cage and locating the pump laser a reasonable distance from the LEI setup.

### 6.5.3 EIE Interferences

EIEs are the most significant source of interferences for LEI. EIEs cause an increase in the level of flame background, induce space charge effects, and may cause a variety of broadband spectral interferences, such as wing excitation of alkali metals, absorption of laser light by molecules that contain alkaline earth metals, and photoionization. The widespread use of LEI for quantitative analysis has been limited by interferences caused by EIEs. Methods to correct for some of these interferences were described above, and other methods are discussed in Sections 6.6.1. and 6.6.6.

## 6.6  Recent Developments in Analytical LEI

This section describes major developments in analytical LEI since the last major review[1] of the topic, covering the period from 1989 to the present. As in earlier work, the majority of research has focused upon LEI in an air-acetylene flame, including several theoretical and analytical papers. Two papers discussed graphite furnace LEI, and three discussed the use of various plasmas as atom cells. Two hyphonated LEI techniques were investigated: electrothermal vaporization (ETV) into a flame, and LEI as a selective ion source for mass spectrometry. Three papers described the coupling of LEI to liquid chromatography (LC) or flow injection (FI) systems. Two of these papers separated the analyte from its matrix by LC or FI, while the third considered speciation of lead. LEI in a glow discharge (GD) source was employed for the determination of uranium in three reports.

### 6.6.1  Flame LEI

#### 6.6.1.1  Fundamental Studies

Axner and coworkers published a series of papers[16–20] that considered fundamental aspects of analytical LEI in air-acetylene flames. Axner and Sjöström[16] compared the sensitivity of two-step LEI to photoionization by investigation of the signal sizes of the methods. They concluded that if excitation by two-step LEI is performed to a level within 3000 cm$^{-1}$ of the ionization limit, this method is several orders of magnitude more sensitive than photoionization techniques. The greater sensitivity of LEI was attributed to greater transition probabilities of high-level bound states compared to those for photoionization, and the reasonably high efficiency of ionization from these bound states (10–100%). In addition, the high laser pulse energies required for photoionization are likely to produce background signals from molecules (e.g., NO) or EIEs.

Axner[17] presented a general theory for the evaluation of the most sensitive one-step LEI transitions for a series of elements by derivation of an expression for the LEI signal strength. In an air-acetylene flame, the model predicted optimum sensitivity would be obtained with excitation to an energy level 3000 cm$^{-1}$ from the ionization limit. The validity of the theory was investigated by comparison of the LEI sensitivities for a series of transitions near the ionization limit of sodium and lithium (Figure 6.5). Maximum sensitivity was obtained with states between 4000 and 4400 cm$^{-1}$ demonstrating good agreement between the theoretical prediction and the experimental results. The general result was then employed to predict the most sensitive transition for Group 1A, 2A, and 3A elements.

Axner et al.[18] reported a technique to reduce spectral interferences caused by sodium by the use of one laser system to ionize sodium atoms prior to determination of the analyte with a second laser system (the probe laser). This method was designated "laser preionization." The preionization beam was directed 4 mm below the probe beam, and the latter was delayed with respect to the former by 400 μs. Various preionization schemes and the excitation used for the probe beam are shown in Figure 6.6, along with the resulting decrease in sodium signal in the probe beam. The use of one-step LEI with photoionization caused a 13% reduction in the sodium signal, while two-step LEI caused a reduction of the sodium signal by 80%, and two-step LEI with photoionization reduced the sodium signal by 83%. Similar experiments were then conducted in which the probe laser was used to detect magnesium. Magnesium was determined in a solution containing a 100-fold excess of sodium, and complete recovery of the magnesium signal was reported with a decrease in the background sodium signal by a factor of five. The authors concluded that this procedure was satisfactory for the reduction of spectral interferences, although the experimental setup was too complicated for routine analysis.

Axner and coworkers[19] investigated the source of sodium-induced spectral interferences by scanning the laser between 220 and 410 nm. The type of interference was shown to be dependent upon the excitation wavelength; between 220 and 240 nm, a sodium signal was caused by photoionization; between 240 and 300 nm, the background was induced by wing excitation of 3s-$np$ ($n \geq 5$) transitions; no definite results were obtained between 300 and 320 nm; between 320 and 340, wing excitation of the 3s-4p transition was responsible for the background signal: and from 360 to 410 nm, the sodium LEI signal was produced by photoionization of the thermally populated 3p state. A graph of the

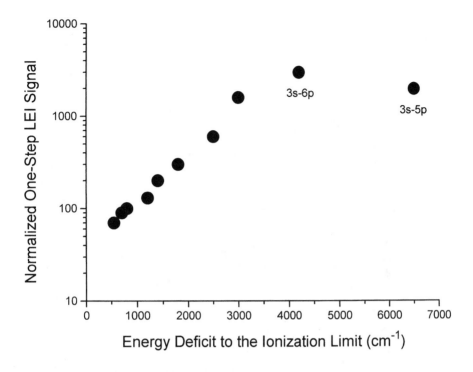

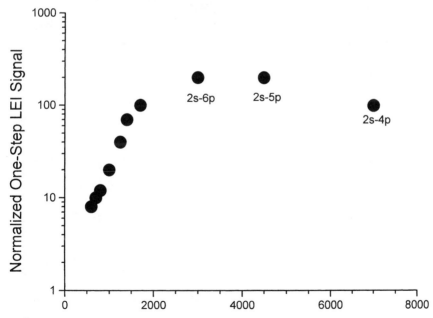

**Figure 6.5** Unsaturated exposure-normalized experimental one-step LEI signal strengths as functions of energy difference between the ionization limit and the state to which the atoms were excited for (**a**) sodium and (**b**) lithium. Adapted with permission from reference 17.

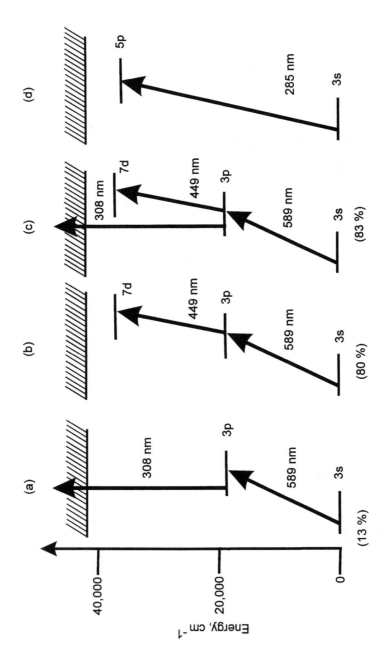

**Figure 6.6** Excitation schemes used for preionization and excitation with the probe beam: (**a**) one-step excitation with photoionization, (**b**) resonant two-step LEI excitation, (**c**) two-step excitation together with photoionization, and (**d**) one-step excitation for the probe beam. Adapted with permission from reference 18.

ratio of LEI signals at 227.66 nm of a potential analyte (bismuth) to that of sodium versus laser pulse energy is shown in Figure 6.7. As the laser power was increased, the ratio of signals decreased because the bismuth transition was saturated at relatively low pulse energies, while the sodium photoionization transition was not. Figure 6.7 implies that optimum sensitivity is obtained at very small pulse energies, which is misleading because other sources of noise dominate under these conditions. A reasonable compromise is to use sufficient laser energy to just saturate the analyte transition, although Axner and co-workers suggest that optimization should be performed for each LEI measurement.

Axner and Sjöström[20] investigated the contribution of "scattered" laser light to excite atoms outside the flame probe volume and produce an LEI signal. When high laser intensities were employed, more than 90% of the ionized atoms were shown to lie outside of the probe volume. The origin of the scatter was not determined, but experiments showed that the scattered light was not due to atomic fluorescence from atoms in the probe volume. In addition, the LEI signal increased as the flame was made more rich, suggesting scatter was caused by constituents that are present in higher concentrations in rich flames. Third, the

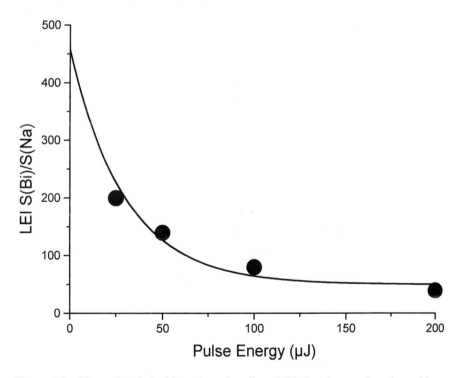

**Figure 6.7** The ratio of the bismuth and sodium LEI signals as a function of laser pulse energy at the bismuth transition at 227.66 nm. The solid line represents the fit of the theoretical treatment to the experimental data. Adapted with permission from reference 19.

signal was independent of laser pulse energy, indicating some type of saturable process.

### 6.6.1.2 Instrumental Developments

Szabo and coworkers[21] described the use of a coil-shaped, water-cooled immersed cathode (Figure 6.8) in order to reduce EIE-induced matrix effects. A series of stainless steel coiled electrodes with spacings of 1,2,4, and 6 mm were constructed and compared to conventional water-cooled cathodes. The laser beam was directed either below or through the coil. The various electrode systems were compared for the determination of indium in the presence and absence of 300 μg/mL sodium. The optimum arrangement was the 4-mm coiled cathode with the laser beam directed through the electrode. The sensitivity of the setup was within a factor of two of a reference water-cooled electrode, and no matrix effects were observed up to a sodium concentration of 2000 μg/mL.

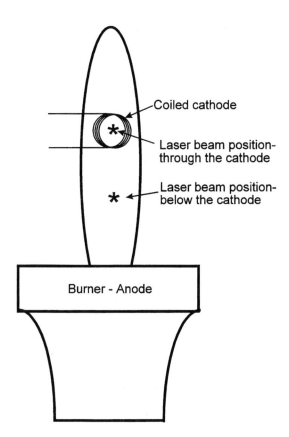

**Figure 6.8** Schematic diagram of the coiled cathode for LEI. Taken with permission from reference 21.

It should be pointed out that this work was comparative in nature, and hence, optimum sensitivity was not obtained. The detection limit reported here with the reference electrode was several orders of magnitude higher than the best value for indium.[1]

### 6.6.1.3 Analytical Results and Applications

Several recent papers have described the determination of elements by flame LEI, and the results are summarized in Table 6.1. Axner and Rubinsztein-Dunlop[12] determined chromium by two-step flame LEI. A relatively high detection limit of 2 ng/mL was obtained, which was attributed to the low efficiency of the air-acetylene flame used to atomize chromium, and contamination from the flame system. Careful optimization of the position and the intensity of the laser beam was shown to significantly reduce interferences caused by high concentrations of sodium.

Miyazaki and Tao[22] determined manganese in water samples by flame LEI with an extraction procedure in order to minimize interferences induced by EIEs. A manganese complex was formed with sodium diethyldithiocarbamate (NaDDC), and subsequently extracted into diisobutyl ketone (DIBK). Optimization of experimental conditions was performed, including laser pulse energy, applied high voltage, and the position of the laser beam. A detection limit of 0.09 ng/mL was obtained with the extraction step, and the calibration curve was linear to 100 ng/mL. The extraction procedure was shown to significantly reduce interferences in water samples. Manganese was determined in ground, river, lake, sea, and tap water with this protocol, and good agreement with the certified or reference values was obtained, with typical relative standard deviations (RSDs) of 2–6%.

Chandola and coworkers[23,24] investigated the determination of sodium and potassium by LEI. For sodium, a detection limit of 0.0001 ng/mL was obtained, which is the lowest value reported to date.[1] A detection limit of 0.1 ng/mL was obtained for potassium, which is an order of magnitude lower than the best

**Table 6.1** Summary of Analytical Applications of Flame LEI, 1989–94

| Reference | Analyte | Pump Laser | Wavelengths (nm) | LOD (ng/mL/ LDR) | Application |
|---|---|---|---|---|---|
| 12 | Cr | Excimer-XeCl | 427/530 | 2/? | None |
| 22 | Mn | Nd:YAG | 279.5 | 0.09 with extraction step for preconcentration/3 | Water |
| 23 | Na | Excimer-XeCl | 578.7 | 0.0001/? | Water |
| 24 | K | Excimer-XeCl | 580.2 | 0.1/2.5 | Water |
| 25 | P | Nd:YAG | 324.6/558.0 | 30/?; used PO molecule | None |

[a] Refers to the laser employed to pump a dye laser for excitation.

reported pulsed laser detection limit and comparable to the best continuous wave laser value. The effect of other alkali metals upon the sodium and potassium LEI signal was investigated. In general, a signal suppression was observed for alkali metals whose ionization potentials were lower than that of the analyte, but not for alkali metals whose values exceeded those of the analyte. Potassium was determined in two water samples, and the results were compared to analyses performed by flame atomic absorption and inductively coupled plasma atomic emission. In general, good agreement was obtained with the three methods.

Turk[25] determined phosphorous using the phosphorous monoxide (PO) molecule in an air-acetylene flame by resonantly enhanced multiphoton ionization (REMPI) and two-step LEI. The two-step LEI measurement had a detection limit of 30 ng/mL, which was a factor of six lower than the best REMPI detection limit (200 ng/mL). The use of high laser pulse energies for REMPI was reported to induce a significant flame background that resulted in the degraded detection limit compared to the LEI value. The high pulse energy required for REMPI also caused a degradation in selectivity compared to two-step LEI. The selectivity values for flame LEI were 2–1000 times higher than those for REMPI, with the exception of sodium.

## 6.6.2 Graphite Furnace LEI

Early papers involving graphite furnace LEI employed relatively old atomizers that have been shown to be unsuitable for the analysis of real samples by atomic absorption spectrometry (AAS).[1] The use of modern furnace technology and a novel furnace design has been investigated for LEI. A summary of the analytical applications of graphite furnace LEI from 1989 to 1994 is given in Table 6.2.

Butcher and coworkers[13,14] investigated probe atomization for LEI in a graphite tube furnace. The goal of this study was to investigate the use of modern furnace technology employed in AAS for LEI in order to evaluate the po-

**Table 6.2**  Summary of Analytical Applications of Graphite Furnace LEI, 1989–94

| Reference | Analyte | Pump Laser[a] | Wavelengths (nm) | LOD (pg/LDR) | Application |
|---|---|---|---|---|---|
| 13.14 | Tl | Excimer-XeCl | 291.8 | 2/3.5 | None |
|  | In |  | 271.0 | 2/3 |  |
|  |  |  | 303.9 | 0.7/3 |  |
|  | Pb |  | 283.3 | 60/3 |  |
|  | Li |  | 274.1 | 1/3 |  |
|  | Fe |  | 302.1 | 50/3 |  |
|  | Mg |  | 285.1 | 10/3 |  |
|  | Mn |  | 279.5 | 30/3 |  |
| 26 | In | Excimer-XeCl | 451/571 | 0.0008/4 | None |
|  | Yb |  | 555/581 | 1/4 |  |

[a]  Refers to the laser employed to pump a dye laser for excitation.

tential of the technique for real sample analysis. A schematic diagram of the detection system employed in this work is shown in Figure 6.9. A graphite probe, whose position was controlled by a stepper motor, was used for sample introduction, as the high voltage electrode, and for collection of the LEI signal. The wall of the furnace was maintained at ground potential and served as the anode. The LEI signal was collected at the probe, passed through a high-pass filter to prevent the high voltage from entering the detection system, preamplified, and sent on to a boxcar integrator.

The timing sequence for this instrument began with the probe inside the graphite tube during the dry and char stages. The laser was activated 2 s before atomization to allow stabilization of the laser output, and baseline was measured for 2 after the initiation of the atomization stage. The probe moved out of the furnace 1 s before atomization, and was reinserted into the furnace 3 s after the initiation of atomization, after the furnace had reached the maximum temperature. This ensured that atomization occurred in a hot, isothermal environment, which has been shown to minimize vapor phase interferences.[10]

Several optimization studies were performed, including the probe voltage, the atomization temperature, and the temporal position of the laser pulse. The probe voltage was optimized at −50 V. Atomization temperature optimization profiles for thallium and iron are shown in Figure 6.10. For thallium, an element of high volatility, the optimum atomization temperature was determined to be 1800°C. Optimization of the atomization temperature for iron, which is of medium volatility, showed that no signal was obtained at temperatures exceeding 2000°C. The disappearance of the signal was attributed to the presence of a

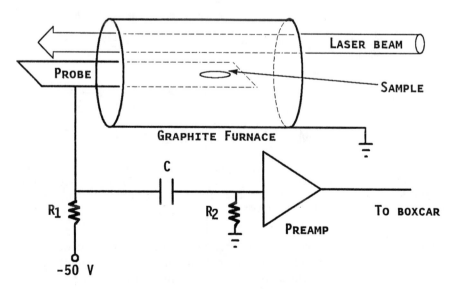

**Figure 6.9** Schematic diagram of the detection system employed for graphite furnace LEI. The movement of the probe is controlled by a stepper motor: $R_1 = R_2 = 22\ k\Omega$; $C = 100$ nF, 100 V. Adapted with permission from reference 13.

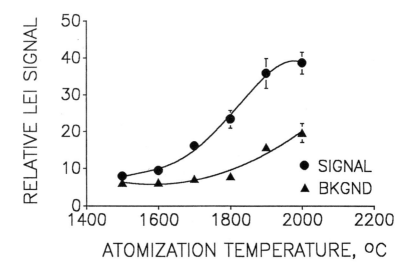

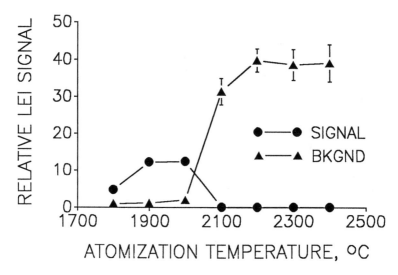

**Figure 6.10** Atomization temperature optimization for graphite furnace LEI (**a**) thallium (100 pg; excitation wavelength, 291.8 nm) and (**b**) iron (500 pg; excitation wavelength, 302.1 nm): (●), signal; and (▲), background. The applied voltage was −50 V. Adapted with permission from reference 13.

large direct current caused by thermionic emission of electrons. It was consequently necessary to employ a maximum atomization temperature of 1900°C. For relatively involatile elements (Fe, Li, Mg, and Mn), a 20-s atomization period was required, and even then, it was not possible to collect the entire signal.

Detection limits between 0.7 and 2 pg were obtained for elements (Tl, In, and

Li) that were excited to within 7000 cm$^{-1}$ of the ionization limit (Table 6.3). The other elements investigated (Fe, Mg, Mn, and Pb) were excited between 24,000 and 31,000 cm$^{-1}$ of the ionization limit, and had correspondingly higher detection limits (10–60 pg). The sensitivity would be significantly improved for the second set of elements by the use of a two-step excitation scheme. The detection limit and sensitivity of graphite furnace LEI[13] are compared to those flame LEI[27] and graphite furnace AAS (GFAAS) in Table 6.3. In general, the sensitivity of graphite furnace LEI, in fC/(ng/mL), was the same, within a factor of five, as that of flame LEI for four of the five elements compared (thallium is the exception), but the detection limits were worse by approximately two orders of magnitude. This indicates that the noise level was approximately 100 times higher in the furnace than the flame, resulting in degraded detection limits in the furnace. Compared to GFAAS, the graphite furnace LEI detection limits were the same (within a factor of five) for elements with favorable one-step transitions (Tl, In, and Li), and 50–100 times worse for elements with less favorable transitions (Fe, Mg, Mn, and Pb).

An investigation of matrix effects showed that sodium caused a severe depression of the indium graphite furnace LEI signal when the mass ratio of sodium to indium exceeded five. Nitric acid did not affect the indium signal, while sulfuric acid affected the graphite furnace LEI measurement only at high concentrations due to oxidation of the graphite furnace. Hydrochloric acid was shown to depress the indium LEI signal due to the formation of the indium chloride, which is a chemical interference not specific to LEI.

Chekalin and Vlasov[26] described a graphite furnace LEI instrument that involved atomization of analyte from a conventional graphite tube furnace and subsequent transport of the analyte through the dosing hole above the furnace. The region above the furnace was irradiated with laser radiation to induce ion-

**Table 6.3**  Comparison of Detection Limits and Sensitivity of Graphite Furnace LEI, Flame LEI, and Graphite Furnace Atomic Absorption Spectrometry (GFAAS)

| Element | Graphite furnace LEI[13,14] | | Flame LEI[27] | | GFAAS |
| | Sensitivity [fC/(ng/mL)] | LOD (ng/mL)$^a$ | Sensitivity [fC/(ng/mL)] | LOD (ng/mL) | LOD (ng/mL) |
|---|---|---|---|---|---|
| Tl | 3 | 0.4 | 18 | 0.006 | 0.5 |
| In | 45 | 0.14 | 50 | 0.001 | – |
| Li | 400 | 0.2 | 190 | 0.005 | 0.1 |
| Fe | 3.5 | 10 | – | – | 0.1 |
| Mg | 4 | 2 | – | – | 0.03 |
| Mn | 4.5 | 6 | 5 | 0.04 | 0.05 |
| Pb | 0.1 | 12 | 0.4 | 0.2 | 0.25 |

$^a$  Using 5 μL sample volume
$^b$  Using 20 μL sample volume
Adapted with permission from reference 13

ization, and an electric field ($-400$ V) was induced between the cathode and the graphite tube at ground potential to collect the charges (Figure 6.11). This approach was investigated to physically separate the collection of the LEI signal from graphite furnace noise, using a scheme similar to Sjöström's T-furnace.[28] Detection limits of 1 pg for ytterbium and 0.8 pg for indium were obtained. Severe vapor-phase interferences would be expected for this atomization system for the analysis of real samples.[10]

### 6.6.3 Plasma LEI

Three plasmas[29–31] have been used as atom cells for LEI, and analytical figures of merit are summarized in Table 6.4. Ng and coworkers[29] did LEI in an extended torch inductively coupled plasma with a continuous wave (CW) argon ion pumped dye laser system. This laser system could only be operated in the visible region, resulting in relatively poor detection limits. A voltage of $-800$ to $-1100$ V was applied across the plasma by use of water-cooled stainless steel electrodes. The laser beam was chopped at 1.8 kHz and the signal was processed with a lock-in amplifier. Detection limits for 6 elements had values between 30 and 920 ng/mL, which are three to five orders of magnitude worse than values

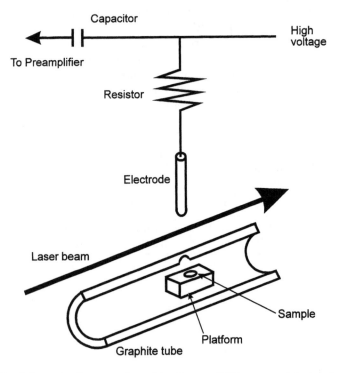

**Figure 6.11**  Schematic diagram of graphite furnace LEI system with signal collection above the graphite tube furnace. Adapted with permission from reference 26.

**Table 6.4** Summary of Analytical Applications of Plasma LEI, 1989–94

| Reference | Analyte | Pump laser[a]/Plasma | Wavelengths (nm) | LOD (ng/mL)/LDR | Application |
|---|---|---|---|---|---|
| 29 | Na | Continuous wave | 589.0 | 30/3 | None |
| | Li | argon ion/inductively | 670.8 | 180/3 | |
| | | coupled plasma (ICP) | 610.3 | 920/3 | |
| | Ca | | 422.7 | 30/? | |
| | Sr | | 460.7 | 810/? | |
| | In | | 451.1 | 220/? | |
| | Ga | | 417.2 | 200/? | |
| 30 | Na | Continuous wave | 589.0 | 10 | None |
| | Yb | argon ion/microwave | | | |
| | | induced plasma (MIP) | | | |

[a] Refers to the laser employed to pump a dye laser for excitation.

obtained by flame LEI. Although the detection limits would certainly be improved with a pulsed laser system, the additional cost, complexity and relatively poor sensitivity of ICP-LEI would appear to eliminate its use for routine analysis.

Lysakowski et al.[30] detected LEI signals for sodium, lithium, and barium in a microwave-induced plasma (MIP) using a CW argon ion pumped dye laser system for excitation and a lock-in amplifier detection system. Circular stainless steel plate electrodes were shown to provide superior calibration curve linearity than tungsten welding rod electrodes. Optimization experiments included the spacing, voltage, and distance of electrodes above the microwave cavity; the microwave power; gas composition; and laser power. Optimum conditions for the determination of sodium were 2-mm spacing between electrodes, 2000 V, 5 mm above the cavity, and 85 W of microwave power; 15% $N_2$/85% Ar as the gas composition and a laser power of 110 mW. A detection limit of 10 ng/mL was obtained for sodium, which is a factor of 100,000 higher than the flame LEI detection limit reported by Chandola et al.[23]

Coche et al.[31] investigated the feasibility of doing laser enhanced ionization in a laser-induced plasma (LIP). The LIP was produced by a nitrogen laser, and a nitrogen-pumped dye laser system was used for LEI excitation. An LEI signal was not obtained with this system, which was explained as due to the characteristics of the LIP. The electron density was equal to the ion density in the majority of the excitation volume, causing zero electric field and a space charge effect (Section 6.5).

### 6.6.4 Electrothermal Vaporization with Flame LEI

Four papers have described the use of electrothermal vaporization for sample introduction with flame LEI (ETV-LEI).[26,32–34] The principal advantages of this scheme are the potential high sensitivity afforded by the direct vaporization of solid and liquid samples by the ETV into the flame for detection by LEI, and the ability to vaporize matrix species before vaporization of the analyte. A summary of analytical figures of merit of ETV-LEI is given in Table 6.5.

Miyazaki and Tao[32] determined thallium in a variety of water samples by ETV-LEI. After optimization of the laser power, cathode potential, ETV temperatures, and flow rate, a detection limit of 0.043 ng/mL thallium was obtained. In order to eliminate interferences caused by EIEs, the analyte was converted into an organic complex and extracted into 2.6-dimethyl-4-heptanone prior to determination. The extraction step was reported to minimize interferences, including those caused by EIEs. The determination of thallium in natural water samples by ETV-LEI was in good agreement with values obtained by inductively coupled plasma atomic emission spectrometry (Table 6.6), although the latter technique did not have adequate sensitivity to determine the lower concentrations of thallium.

Chekalin et al.[26,33] determined four elements by ETV-LEI in four high-purity materials (Table 6.5). The ETV unit was a graphite rod atomizer that was heated to dry and char the sample, and subsequently inserted into the flame to cause vaporization and atomization. Detection limits between 2 and 200 fg were obtained: in the case of copper, the LOD was contamination limited. Results of the analysis by direct solid sampling and a dissolution procedure are shown in Table 6.7, along with detection limits by each method of preparation. Although the accuracy of the measurements was not evaluated, good agreement between the methods of sampling was achieved. The determination of gold in silver nitrate was not successful because of a significant background signal.

Smith and coworkers[34] described the combination of a graphite tube furnace with a miniature flame for ETV-LEI (Figure 6.12). The miniature flame system had advantages of minimizing dilution of the analyte during transport from the furnace and was efficiently illuminated by the Nd:YAG dual dye laser system. Detection limits between 17 and 260 fg were obtained for three elements (Table 6.5), which corresponds to 1.7–26 pg/mL for a 10-μL injection. These detection limits were the same (within a factor of five) as those obtained previously by

**Table 6.5**　Summary of Analytical Applications of ETV-LEI, 1989–94

| Reference | Analyte | Pump Laser[a] | Wavelengths (nm) | LOD (ng/mL)/ LOD (pg)/LDR | Application |
|---|---|---|---|---|---|
| 32 | Tl | Nd:YAG | 276.8 | 0.043/0.86/2.3 | Water |
| 26.33 | In | Excimer-XeCl | 451/571 | 0.00004/0.0004/4 | Na, Cu in $H_3PO_4$ |
| | Na | | 589/449 | 0.00002/0.0002/4 | Cu in Ge |
| | Cu | | 324/453 | 0.002/0.02/4 | In in CdHgTe |
| | Au | | 267/294 | 0.002/0.02/4 | Au in $AgNO_3$ |
| 34 | In | Nd:YAG | 303.9/786.4 | 0.0260/0.260/5 | None |
| | Mg | | 285.2/435/2 | 0.0017/0.017/5 | |
| | Tl | | 377.6/655.6 | 0.0118/0.118/5 | |

[a]　Refers to the laser employed to pump a dye laser for excitation

**Table 6.6**  Determination of Thallium by ETV-LEI and Inductively Coupled Plasma Atomic Emission Spectrometry (ICP-AES) with Solvent Extraction

| Water sample | ETV-LEI (ng/mL) | ICP-AES (ng/mL) |
|---|---|---|
| River A | 270 ± 11.5 | 270 |
| River B | 158 ± 6.5 | 180 |
| Lake A | 0.2 ± 0.06 | Not Detected. |
| Coastal sea water A | 1.1 ± 0.2 | Not Detected. |
| Coastal sea water B | 0.9 ± 0.1 | Not Detected. |
| Hot spring A | 31 ± 1.3 | 33 |
| Hot spring B | 4.1 ± 0.8 | 3.6 |

Adapted with permission from reference 32

two-step flame LEI, and hence the added complexity of the electrothermal vaporization system did not seem to improve detection limits. A potential advantage of this technique, which was not experimentally evaluated, is the use of the ETV system to remove matrix components prior to atomization. These procedures have been commonly employed for ETV-ICP and ETV-ICP-MS.[35] The development of ETV-LEI methods that reduce interferences would be a significant development for the routine use of LEI for real sample analysis.

## 6.6.5  Laser-Enhanced Ionization with Mass Spectrometric Detection

Two papers have described the use of LEI as a selective ion source for mass spectrometry (Table 6.8).[36,37] Yu et al.[36] reported the use of LEI and direct laser ionization (DLI) in an inductively coupled plasma (ICP) with a mass spectrometer (MS) as the detector. This work is distinguished from resonance ionization mass spectrometry (RIMS) because of the use of a conventional atmospheric-

**Table 6.7**  Analysis of Real Samples by ETV-LEI[26,33 a]

| Sample | Analyte | Solution, 10 µL | | Solid, 10 mg | |
|---|---|---|---|---|---|
| | | Concentration (µg/mL) | LOD (pg) | Concentration (µg/mL) | LOD (pg) |
| Ge #1 | Cu | 790 ± 60 | | 710 ± 70 | 5 |
| Ge #2 | | 370 ± 60 | | 300 ± 40 | |
| CdHgTe #1 | In | 32 ± 2 | 0.03 | 34 ± 3 | 0.01 |
| CdHgTe #2 | | 29 ± 2 | | 26 ± 3 | |
| $H_3PO_4$ | Na | 220,000 ± 20,000 | 0.0003 µg/mL | – | – |
| | Cu | 8400 ± 800 | 0.00002 µg/mL | – | – |

[a]  Uncertainties represent ± 1 standard deviation of the mean for ten measurements. LOD, limit of detection.

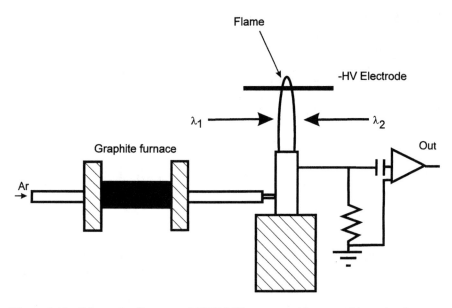

**Figure 6.12** Schematic diagram of ETV-LEI system with a graphite tube furnace. Adapted with permission from reference 34.

pressure atom cell.[38] A schematic diagram of the instrumentation employed for the determination of strontium is shown in Figure 6.13. An ionizing laser beam passed through port A, while a second probe beam passed through port B. The role of the probe beam was to monitor the population of neutral strontium atoms by atomic fluorescence, which was collected by a lens and an optical fiber. The DLI experiment showed an 11% enhancement of the ionized strontium signal compared to plasma-induced ionization background. The weak DLI sig-

**Table 6.8** Applications of LEI as a Source of Ions for Mass Spectrometry, 1989–94

| Reference | Analyte | Pump laser[a] | Wavelengths (nm) | Signal enhancement compared to background ionization | Application |
|---|---|---|---|---|---|
| 36 | Sr | Excimer-XeCl | DLI 460.7/308 | 11%[b] | None |
| | | | 1-step LEI | No signal[b] | |
| | | | 460.7 | 1.7%[b] | |
| | | | 2-step LEI | | |
| | | | 460.7/554.3 | | |
| 37 | Na | Excimer-XeCl | 589.0/498.3 | 35.000%[c] | None |

[a] Refers to the laser employed to pump a dye laser for excitation
[b] Using a 10-μg/mL standard
[c] Using a 500-ng/mL standard

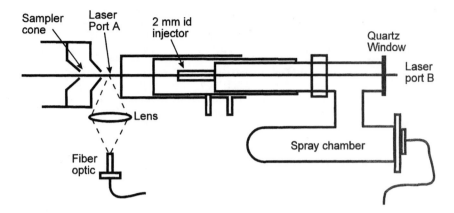

**Figure 6.13** Schematic diagram of instrumentation for LEI-ICP-MS. Adapted with permission from reference 36.

nals were attributed to ion-electron recombination reactions which prevent detection of the majority of the laser-induced charges. No laser-induced enhancement was observed with one-step LEI, and two-step LEI showed an enhancement of 1.7%. The authors concluded that conditions for collisional ionization in the ICP were considerably weaker than in a flame.

Turk and coworkers[37] investigated the determination of Na by use of two-step flame LEI as an ion source for mass spectrometry. The maximum signal was obtained 0.4 ms after the laser pulse, and the halfwidth of the signal was 0.54 ms. Investigation of the position of the laser beam showed that the LEI signal was relatively constant when the distance between the laser beam and the MS interface was between 1.5 and 8 mm. Although analytical figures of merit were not measured for this system, this technique seems to have promise for high-sensitivity analysis.

### 6.6.6 LEI Interfaced with Chromatography and Flow-Injection Techniques

LEI has been interfaced with separation techniques to achieve one of the following goals: (1) reduction of EIE interferences for conventional elemental analysis and (2) speciation of metals. Two publications[15,39] have described the use of separation techniques to reduce interferences, while one has reported on the speciation of lead.[40]

#### 6.6.6.1 LEI Interfaced with Chromatography and Flow-Injection to Minimize Interferences

Turk and Kingston[15] described the use of chelation liquid chromatography (LC) to separate EIEs from samples before LEI analysis (Table 6.9). A series of

**Table 6.9**   Summary of Chromatographic and Flow Injection-LEI, 1989–94

| Reference | Analyte | Pump laser[a] | Wavelengths (nm) | LOD (ng/mL)/ LDR | Application |
|---|---|---|---|---|---|
| 15 | Cd | Nd:YAG | 228.8/466.2 | ?/? | Water, bovine |
|    | Co |        | 252.1/591.7 | ?/? | serum, river |
|    | Cu |        | 324.8/521.8 | ?/? | sediment, food, |
|    | Mn |        | 279.5/521.2 | ?/? | apple leaves, |
|    |    |        | 279.8       | ?/? | peach leaves. |
|    |    |        | 279.5/521.5 | ?/? | |
|    | Ni |        | 300.2/561.5 | ?/? | |
|    | Pb |        | ?/?         | ?/? | |
| 39 | In | Nd:YAG | 325.6       | ?/? | None |

[a]   Refers to the laser employed to pump a dye laser for excitation

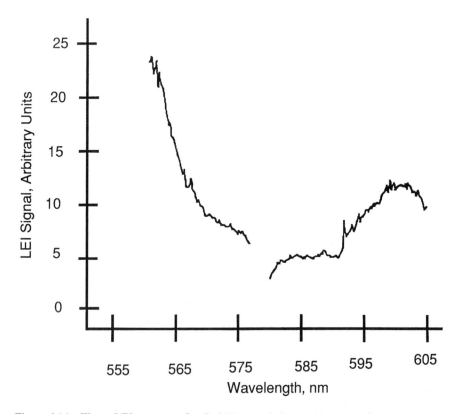

**Figure 6.14**   Flame LEI spectrum for CaOH, recorded over the range of three laser dyes while aspirating a solution of 10 µg/mL Ca. Adapted with permission from reference 15.

standard reference materials (SRMs) were analyzed for Cd, Co, Cu, Mn, Ni, and Pb to evaluate the accuracy of the technique. The samples were acid-digested, and buffered to a pH range of 5.3–5.6. The EIEs (alkali and alkaline earth metals) were eluted from a 4-mL slurry of Chelex-100 resin in a polypropylene column by 1 M ammonium acetate and discarded, followed by elution of the analytes with nitric acid. The preconcentrated solution was then analyzed by two-step LEI.

Several examples of broadband spectral interferences were demonstrated. Figure 6.14 shows an interference caused by the formation of CaOH. The presence of 100 μg/mL sodium caused a large background signal which significantly degraded the signal-to-noise ratio for nickel (Figure 6.15).

Turk and Kingston[15] used LC-LEI to determine six elements in a variety of standard reference materials (SRMs), including water, bovine serum, river sediment, total diet, and leaves. Figures 6.16 and 6.17 show the ability of the sep-

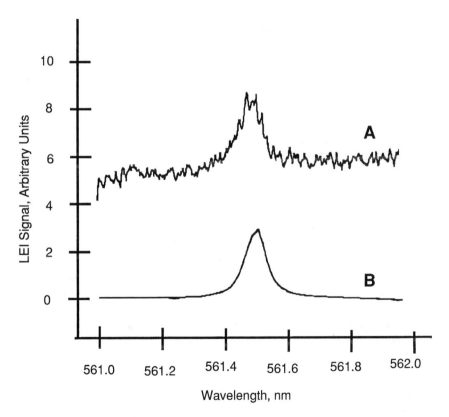

**Figure 6.15** Two-step LEI spectra for 100 ng/mL nickel: (**a**) with and (**b**) without 100 μg/mL sodium present in the solution. The first-step transition is the nickel transition at 300.249 nm. Adapted with permission from reference 15.

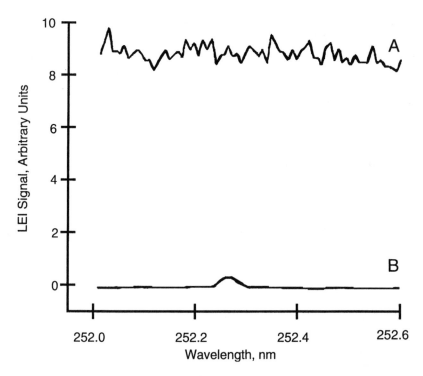

**Figure 6.16** First-step excitation LEI spectra near the cobalt resonance line for SRM 1643b, trace elements in water: (**a**) with and (**b**) with the chromatography step. The second-step laser was held at the cobalt transition at 591.680 nm. Adapted with permission from reference 15.

aration procedure to reduce the background signal in a water sample for cobalt and cadmium, respectively. The results of the analyses are shown in Table 6.10, and, in general, good agreement between the LEI and certified values was obtained. The combination of liquid chromatography with LEI appears to be a viable method for the determination of trace elements in complex samples.

Wang and Lin[39] investigated the use of flow injection (FI) with flame LEI of indium to reduce interferences caused by sodium. A single-channel FI system was used to inject a segment of distilled water into the stream of sample solution. The authors reported a reduction of interferences with the FI setup, although the potential of this system for real sample analysis was not evaluated.

### 6.6.6.2 Speciation by Flame LEI

Epler et al.[40] combined high performance liquid chromatography (LC) with flame LEI to do speciation of organolead compounds in an oyster tissue SRM (Table 6.11). A chromatogram of four compounds is shown in Figure 6.18. A hexane extraction and a tetramethylammonium hydroxide digestion were used as sample preparation procedures for spiked and unspiked samples. The recov-

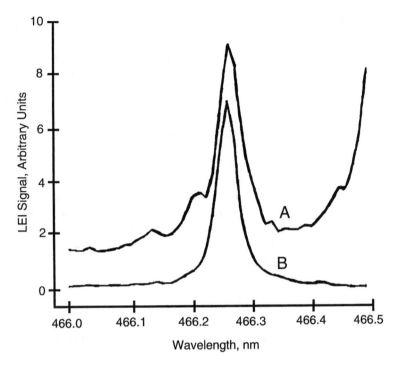

**Figure 6.17**   Second-step excitation LEI spectra near the cadmium resonance line for SRM 1643b, Trace Elements in Water: (**a**) without and (**b**) with the chromatography step. The first-step laser was held at the cadmium transition at 228.802 nm. Adapted with permission from reference 15.

**Table 6.10**   Results of Analyses by LC-LEI[15]

| Sample | Element (replicates) | LEI result (ng/g) | Certified concentration (ng/g) |
|---|---|---|---|
| Water (SRM 1643b) | Cd ($n = 9$) | 20.1 ± 0.47 | 20 ± 1 |
| | Co ($n = 8$) | 26.0 ± 0.22 | 26 ± 1 |
| | Cu ($n = 7$) | 21.8 ± 0.33 | 21.9 ± 0.4 |
| | Mn ($n = 7$) | 30.1 ± 0.78 | 28 ± 2 |
| | Ni ($n = 12$) | 46.6 ± 0.58 | 49 ± 3 |
| | Pb ($n = 7$) | 23.9 ± 0.66 | 23.7 ± 0.7 |
| Bovine serum (SRM 1598) | Mn ($n = 12$) | 3.87 ± 0.13 | 3.78 ± 0.32 |
| | Ni ($n = 4$) | 0.76 ± 0.27 | Not Available |
| Buffalo River Sediment (SRM 2704) | Ni ($n = 18$) | 45.7 ± 1.6 | 44.1 ± 3.0 |
| | Pb ($n = 16$) | 154.4 ± 7.0 | 161 ± 17 |
| Total Diet (SRM 1548) | Cu ($n = 12$) | 2.43 ± 0.10 | Not Available |
| | Mn ($n = 24$) | 5.04 ± 0.14 | Not Available |
| | Ni ($n = 12$) | 0.41 ± 0.08 | Not Available |
| Apple leaves (SRM 1515) | Mn ($n = 12$) | 54.9 ± 1.1 | Not Available |
| | Ni ($n = 12$) | 0.97 ± 0.06 | Not Available |
| Peach leaves (SRM 1547) | Mn ($n = 12$) | 98.4 ± 1.2 | Not Available |
| | Ni ($n = 12$) | 0.74 ± 0.04 | Not Available |

**Table 6.11** Speciation by flame LEI, 1989–94

| Reference | Analyte | Pump laser[a] | Wavelengths (nm) | LOD (ng/mL)/LOD (pg)/LDR | Application |
|---|---|---|---|---|---|
| 40 | Organolead compounds | Nd:YAG | 283.3/600.2 | 0.09/20/? | Oyster tissue |

[a] Refers to the laser employed to pump a dye laser for excitation

ery of a 100-µg/mL spike of organoleads was only 0.5% with the extraction method. The digestion procedure had higher recovery values, but the values varied from 24 to 48% for the four organolead compounds determined. The authors suggested that, although further development of the sample preparation procedures was required for the oyster tissue analysis, the use of LC-LEI for the speciation of organolead compounds was demonstrated.

### 6.6.7 Glow Discharge LEI

Three publications have described the use of glow discharge LEI (GD-LEI) for the determination of uranium (Table 6.12). Lipert and coworkers[41] demon-

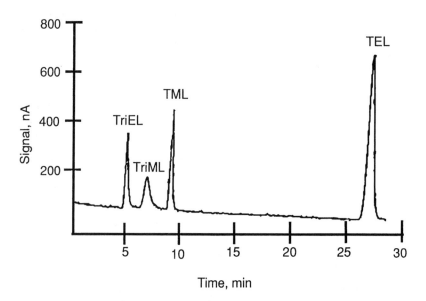

**Figure 6.18** Chromatogram of LC separation of four organolead compounds, equivalent to 500 ng/mL each, in 70% methanol–30% water. TriEL, triethyllead: TriML, trimethyllead; TML, tetramethyllead; TEL, tetraethyllead. The eluent is 9 + 1 methanol-water containing 0.01 mol/L ammonium acetate at a flow rate of 1 mL/min. The sample size was 20 µL. Adapted with permission from reference 40.

**Table 6.12** Glow Discharge LEI, 1989–95

| Reference | Analyte | Pump laser[a] | Wavelengths (nm) | LOD (ng/mL)/LOD (pg)/LDR | Application |
|---|---|---|---|---|---|
| 41 | U | GaAlAs diode | 778.4 | Not applicable | None |
| 42, 43 | [235]U | Ar[+] Ti:sapphire ring; GaAlAs diode | Various 776.2, 831.8 | ?? | [235]U enriched uranium |

[a] Refers to the laser employed to pump a dye laser for excitation

strated the feasibility of GD-LEI for the detection of uranium in a hollow cathode lamp using a GaAlAs diode laser as the source. Barshick et al.[42,43] developed a GD-LEI system for the determination of [235]U using a continuous wave argon ion–pumped titanium:sapphire ring laser and a GaAlAs diode laser. A demountable cathode allowed analysis of real samples with the system. The ultimate goal of this project was the development of a field instrument for isotopic analysis of uranium with the diode laser system. Preliminary work was performed with the argon ion laser system because of its greater ease of use. LEI spectra for uranium are shown in Figure 6.19, demonstrating the applicability of both laser systems to determine uranium isotopes. The system was evaluated by analysis of a [235]U-enriched uranium sample with the diode laser system. Internal precision (run-to-run) was better than 1% relative standard deviation (RSD) and external precision (sample-to-sample) was approximately 3% RSD. The accuracy and precision were reported to be satisfactory for screening applications.

## 6.7 Summary

LEI is a very sensitive method of elemental analysis using the flame as the atom cell. A summary of flame LEI detection limits obtained to date is given in Table 6.13, and compared to detection limits of the most sensitive commercially available atomic spectrometry method, inductively coupled plasma mass spectrometry (ICP-MS). In general, for most elements whose LEI conditions have been carefully optimized (Ca, Cu, In, K, Li, Mg, Na, Pb, and Sr), LEI detection limits are the same or better than those obtained by ICP-MS. On the other hand, a number of elements have been determined using nonoptimum conditions, and for many their LEI detection limits can be improved by one or more of the following approaches: (1) the use of a more sensitive one-step transition (As, Bi, Fe, Si, Sb); (2) the use of two-step LEI (As, Bi, Fe, Ga, Lu, Sb, Si, Ti, Tm, V, W, Y); and (3) the use of the most sensitive flame for a particular analyte (e.g., Ba, Cr, Ga, P, W, and Yb are involatile elements that are probably most sensitively determined in a nitrous oxide–acetylene flame).[10] Optimization of the laser conditions (with either one-step or two-step excitation) and the atom cell

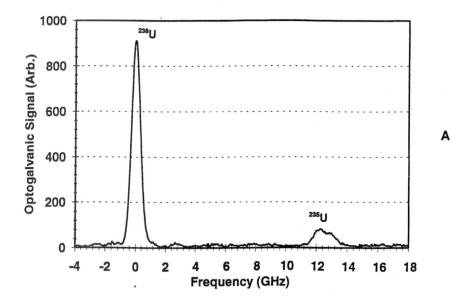

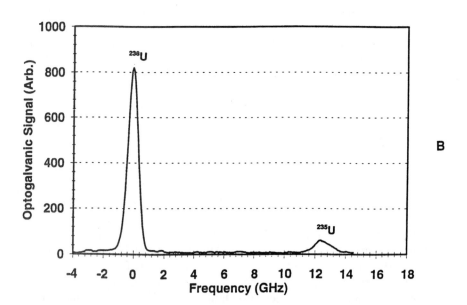

**Figure 6.19**    Uranium LEI spectra recorded for the 776.19 nm line: (**a**) titanium:sapphire laser at approximately 300 W/cm$^2$ and (**b**) diode laser at approximately 400 W/cm$^2$. The sample was the same in both cases, enriched uranium metal (9.97% $^{235}$U). The discharge conditions were the same in both cases, 800 Pa argon, 20 mA, and 300 V. Adapted with permission from reference 42.

**Table 6.13** Flame LEI and Inductively Coupled Plasma Mass Spectrometry (ICP-MS) Limits of Detection (LOD).

| Element | Wavelengths (nm) | Pump laser | Flame | LOD (ng/mL)[6] | ICP-MS (ng/mL)[44] |
|---------|------------------|------------|-------|----------------|--------------------|
| Ag | 328.1/421.1 | Excimer | Air–Acetylene | 0.07 | 0.005 |
| Al | 309.3 | Flashlamp | $N_2O$–acetylene | 0.2 | 0.015 |
| As | 278.0[a] | Excimer | Air–acetylene | 3000 | 0.031 |
| Au | 242.8/479.3 | Nd:YAG | Air–acetylene | 1 | 0.005 |
| Ba | 307.2 | Flashlamp | Air–acetylene[b] | 0.2 | 0.006 |
| Bi | 227.7[a] | Excimer | Air–acetylene | 0.2[45] | 0.004 |
| Ca | 227.5 | Excimer | Air–acetylene | 0.006[45] | 10[46] |
| Cd | 228.8/466.2 | Nd:YAG | Air–acetylene | 0.1 | 0.012 |
| Co | 252.1/591.7 | Nd:YAG | Air–acetylene | 0.08 | 0.005 |
| Cr | 252 | Excimer | Air–acetylene[b] | 0.2 | 0.04 |
| Cs | 455.5 | Nitrogen | Propane–butane–air | 0.004 | 0.002 |
| Cu | 324.8/453.1 | Nd:YAG | Air–acetylene | 0.7 | 0.5[46] |
| Fe | 302.1[a] | Excimer | Air–acetylene | 0.08 | 0.58 |
| Ga | 241.9 | Excimer | Air–acetylene[b] | 0.03[45] | 0.004 |
| In | 451.1/571.0 | Nitrogen | Air–acetylene | 0.0009 | 0.002 |
| K | 580.2 | Excimer | Air–acetylene | 0.1[24] | 15[46] |
| Li | 670.8/460.3 | Excimer | Air–acetylene | 0.0003 | 0.027 |
| Lu | 308.2 | Flashlamp | $N_2O$–acetylene | 0.2 | Not Available |
| Mg | 285.2 | Excimer | Air–acetylene | 0.003 | 0.018 |
| Mn | 279.5/521.5 | Excimer | Air–acetylene | 0.02 | 0.006 |
| Mo | 319.4 | Flashlamp | $N_2O$–acetylene | 10 | 0.006 |
| Na | 578.7/578.7 | Excimer | Air–acetylene | 0.0001[23] | 0.11 |
| Ni | 229.0 | Excimer | Air–acetylene | 0.02[45] | 0.013 |
| P | 324.6/558.0 | Nd:YAG | Air–acetylene[b] | 30[25] | Not Available |
| Pb | 283.3/600.2 | Nd:YAG | Air–acetylene | 0.09 | 0.5[46] |
| Rb | 780.0? | Nitrogen | Propane–butane–air | 0.1 | 0.005 |
| Sb | 287.8[a] | Excimer | Air–acetylene | 50 | 0.012 |
| Sc | 301.9 | Flashlamp | $N_2O$–acetylene | 0.2 | 0.015 |
| Si | 288.2[a] | Flashlamp | $N_2O$–acetylene | 40 | Not Available |
| Sn | 284.0/597.0 | Nd:YAG | Air–hydrogen | 0.3 | 0.01 |
| Sr | 230.7 | Excimer | Air–acetylene | 0.003[45] | 0.003 |
| Ti | 320.0 | Flashlamp | $N_2O$–acetylene | 1 | 0.011 |
| Tl | 291.8/377.6 | Excimer | Air–acetylene | 0.008 | 0.003 |
| Tm | 297.3 | Flashlamp | $N_2O$–acetylene | 200 | Not Available |
| V | 318.4 | Flashlamp | $N_2O$–acetylene | 0.9 | 0.008 |
| W | 283.1 | Excimer | Air–acetylene[b] | 300 | 0.007 |
| Y | 298.4 | Flashlamp | $N_2O$–acetylene | 10 | 0.004 |
| Yb | 267.2 | Excimer | Air–acetylene[b] | 2 | Not Available |
| Zn | 213.9/396.6 | Nd:YAG | Air–acetylene | 1 | 0.035 |

[a] Nonoptimum excitation wavelength

[b] Nonoptimum flame

LEI detection limits were adapted from reference 5 unless indicated otherwise will improve detection limits for many of these elements by one to three orders of magnitude. On the other hand, there are a handful of elements with high ionization potentials (e.g., Cd and Zn) that are not very sensitive to analysis by LEI and for whom little improvement is expected.

A major limitation for the use of flame LEI for real sample analysis is the presence of interferences, primarily due to easily ionized elements. The LC-LEI system described by Turk and Kingston[15] appears to be a successful approach to minimizing most interferences.

Assuming that LC is capable of removing interferences, the major impediments to wider applicability of flame LEI are the unreliability of the laser systems and its single-element capability. As discussed above in Section 6.4.1, laser systems employed to date are insufficiently reliable and easy to use for routine analysis. With current laser systems, in order to obtain optimum sensitivity, changing elements usually means replacement of the dye in the dye laser, which is a very time-consuming, messy procedure. The availability of a relatively inexpensive, reliable, easy-to-use system may allow more widespread use of LEI.

To date, other atom cells, such as graphite furnaces and plasmas, have not demonstrated sufficiently high sensitivity for LEI compared to commercially available absorption and emission methods to justify the complexity of the laser system. One area that may prove analytically useful is the combination of flame LEI with mass spectrometry, although more development needs to be done in this area. Another application is the development of LEI instrumentation for the determination of specific elements. For example, Barshick et al.[42,43] constructed a field LEI glow discharge instrument and demonstrated its application for the determination of uranium isotopes.

## Acknowledgment

The author would like to thank Christopher Barshick for providing original figures for this chapter.

## References

1. Green, R.B.; Seltzer, M.D. *Adv. Atom. Spectrom.* **1992,** *1,* 37. Edited Sneddon, J., JAI Press, Inc., Greenwich, CT.

2. Axner, O.; Rubinsztein-Dunlop, H. *Spectrochim. Acta* **1989,** *44B,* 835.

3. Axner, O.; Rubinsztein-Dunlop, H.; Sjöström, S. *Mikrochim. Acta* **1989,** *III,* 197.

4. Kuzyakov, Yu. Ya.; Zorov, N.B. *CRC Crit. Rev. Anal. Chem.* **1988,** *20,* 221.

5. Turk, G.C. *J. Anal. Atom. Spectrom.* **1987,** *2,* 573.

6. Sjöström, S.; Mauchien, P. *Spectrochim. Acta Rev.* **1993,** *15,* 153.

7. Niemax, K. *Fresenius Z. Anal. Chem.* **1990,** *337,* 551.

8. Omenetto, N. *Appl. Phys. B* **1988**, *46*, 209.

9. Travis, J.C.; Turk, G.C.; Devoe, J.R.; Schenck, P.K.; van Dijk, C.A. *Prog. Anal. Atom. Spectrosc.* **1984**, *7*, 199.

10. *Atomic Absorption Spectrometry: Theory, Design, and Applications;* Haswell, S.J., Ed.; Elsevier: Amsterdam, 1991.

11. Butcher, D.J.; Dougherty, J.P.; Preli, F.R.; Walton, A.P.; Irwin, R.L.; Michel, R.G. *J. Anal. Atom. Spectrom.* **1988**, *3*, 1059.

12. Axner, O.; Rubinsztein-Dunlop, H. *Appl. Opt.* **1993**, *32*, 867.

13. Butcher, D.J.; Irwin, R.L.; Sjöström, S.; Walton, A.P.; Michel, R.G. *Spectrochim. Acta* **1991**, *46*, 9.

14. Butcher, D.J.; Ph.D. Dissertation, University of Connecticut, 1990.

15. Turk, G.C.; Kingston, H.M. *J. Anal. Atom. Spectrom.* **1990**, *5*, 595.

16. Axner, O.; Sjöström, S. *Appl. Spectrosc.* **1990**, *44*, 144.

17. Axner, O. *Spectrochim. Acta* **1990**, *45B*, 561.

18. Axner, O.; Norberg, M.; Persson, M.; Rubinsztein-Dunlop, H.; *Appl. Spectrosc.* **1990**, *44*, 1117.

19. Axner, O.; Norberg, M.; Rubinsztein-Dunlop, H. *Appl. Spectrosc.* **1990**, *44*, 1124.

20. Axner, O.; Sjöström, S. *Appl. Spectrosc.* **1990**, *44*, 864.

21. Szabo, N.J.; Latz, H.W.; Petrucci, G.A.; Winefordner, J.D. *Anal. Chem.* **1991**, *63*, 704.

22. Miyazaki, A.; Tao, H. *J. Anal. Atom. Spectrom* **1991**, *6*, 173.

23. Chandola, L.C.; Khanna, P.P.; Razvi, M.A.N. *Anal. Lett.* **1991**, *24*, 1685.

24. Chandola, L.C.; Khanna, P.P. *Spectrochim. Acta* **1992**, *48A*, 1547.

25. Turk, G.C. *Anal. Chem.* **1991**, *63*, 1607.

26. Chekalin, N.V.; Vlasov, I.I.; *J. Anal. Atom. Spectrom.* **1992**, *7*, 225.

27. Axner, O.; Magnusson, I.; Peterson, J.; Sjöström, S. *Appl. Spectrosc.* **1987**, *41*, 19.

28. Sjöström, S.; Magnusson, I.; Lejon, M.; Rubinsztein-Dunlop, H.; *Anal. Chem.* **1988**, *60*, 1629.

29. Ng, K.C.; Angebranndt, M.J.; Winefordner, J.D. *Anal. Chem.* **1990**, *62*, 2506.

30. Lysakowki, R.S.; Dessy, R.E.; Long, G.L. *Appl. Spectrosc.* **1989**, *43*, 1139.

31. Coche, M.; Berthoud, T.; Mauchien, P.; Camus, P. *Appl. Spectrosc.* **1989**, *43*, 646.

32. Miyazaki, A.; Tao, H. *Anal. Sci.* **1991**, *7 (Supplement)*, 1053.

33. Chekalin, N.V.; Pavlutskaya, V.I.; Vlasov, I.I. *Spectrochim. Acta* **1991**, *46B*, 1701.

34. Smith, B.W.; Petrucci, G.A.; Badini, R.G.; Winefordner, J.D. *Anal. Chem.* **1993**, *65*, 118.

35. *Inductively Coupled Plasmas in Analytical Atomic Spectrometry*, 2nd ed.; Montaser, A.; Golightly, D.W., Eds., VCH: Weinheim, 1992.

36. Yu, L.; Koirtyohann, S.R.; Turk, G.C.; Salit, M.L. *J. Anal. Atom. Spectrom.* **1994**, *9*, 997.

37. Turk, G.C.; Yu, L.; Koirtyohann, S.R. *Spectrochim. Acta.* **1994**, *49B*, 1537.

38. Young, J.P.; Shaw, R.W.; Smith, D.H.; *Anal. Chem.* **1989**, *61*, 1271A.

39. Wang, S.-C.; Lin, K.-C. *Anal. Chem.* **1994**, *66*, 2180.

40. Epler, K.S.; O'Haver, T.C.; Turk, G.C. *J. Anal. Atom. Spectrom.* **1994**, *9*, 79.

41. Lippert, R.J.; Lee, S.C.; Edelson, M.C.; *Appl. Spectrosc.* **1992,** *46,* 1307.

42. Barshick, C.M.; Shaw, R.W.; Young, J.P.; Ramsey, J.M. *Anal. Chem.* **1994,** *66,* 4154.

43. Barshick, C.M.; Shaw, R.W.; Young, J.P.; Ramsey, J.M. *Anal. Chem.* **1995,** *67,* 3814.

44. Takahashi, J.; Hara, R. *Anal. Sci.* **1988,** *4,* 331.

45. Axner, O.; Magnusson, I.; *Physica Scripta* **1985,** *31,* 587.

46. Commercial Literature, Fisons Instruments, Beverly, MA.

# Index